Dreamland

Dreamland

ADVENTURES IN THE STRANGE SCIENCE OF SLEEP

David K. Randall

W. W. NORTON & COMPANY

New York • London

Printed in the United States of America
First Edition

For information about special discounts for bulk purchases, please contact
W. W. Norton Special Sales at specialsales@wwnorton.com or 800-233-4830

Manufacturing by Courier Westford
Book design by Chris Welch
Production manager: Anna Oler

Library of Congress Cataloging-in-Publication Data

Randall, David K.
Dreamland : adventures in the strange science of sleep / David K. Randall.
p. cm.
Includes bibliographical references.
ISBN 978-0-393-08020-9 (hardcover)
1. Sleep. 2. Dreams. I. Title.
QP425.R29 2012
612.8'21—dc23

2012014932

W. W. Norton & Company, Inc.
500 Fifth Avenue, New York, N.Y. 10110
www.wwnorton.com

W. W. Norton & Company Ltd.
Castle House, 75/76 Wells Street, London W1T 3QT

1 2 3 4 5 6 7 8 9 0

For Megan

That we are not much sicker and much madder than
we are is due exclusively to that most blessed and
blessing of all natural graces, sleep.

—ALDOUS HUXLEY

Contents

Dreamland

1

I Know What You Did Last Night

One night, not long ago, a man found himself collapsed in a hallway, clutching his leg like a wounded bear. As his curses and howls echoed through the walls of his apartment on an otherwise silent Tuesday night, a thought passed across his brain: something had gone wrong. It was after midnight. He was not supposed to be in this position, on his back on a hardwood floor, and he was definitely not supposed to be in this much pain. He lay there, hurt and confused, not knowing what had happened, since his last memory was laying his head down on a pillow in the bedroom thirty feet away.

That man was me. It had never occurred to me, before that moment, that falling asleep could lead to injury. But there I was, in my boxers, piecing together the last few hours of my life like a disheveled detective who came late to a crime scene. Three things were immediately clear: 1) I had crashed into a wall in my apartment while sleepwalking; 2) I don't sleepwalk with my arms out in front of me like a zombie, which is a pity because 3) sleepwalking into a wall really hurts.

This was the first time that I sleepwalked, or at least the first time that I did it so badly that I ran into something. But sleep had long been a less-than-peaceful part of my life. As a child, I often fell asleep with my eyes open, a condition that unnerved my parents and spooked friends at sleepovers. When I was in college, I unknowingly entertained my roommates while I was sleeping by sitting up and yelling things like "Man the barricades, the bacon is coming!" Now, as a married adult, my wife is treated to a nightly show that can include talking, singing, laughing, humming, giggling, grunting, bouncing, and kicking. She handles all of this by putting in earplugs after we say good night and moving to the far side of our oversized bed, a purchase she insisted on after being on the receiving end of one well-placed kick.

The talking and kicking she could deal with. But she put her foot down once I became mobile. After a few days of limping around and hoping no one would ask why, I found myself in a sleep lab at a New York hospital. The room was decorated like a hotel room in Florida, complete with a pink watercolor painting of a palm tree hanging above what at first glance looked like a headboard. A deeper inspection revealed that it was a piece of wood bolted to the wall above a standard-issue hospital bed.

The walls were painted cream, and one of the last working television-VCR combinations sat on top of a desk in the corner. Medical equipment shared space with a white beach shell on a wooden nightstand.

My brain waves were to be recorded as I slept that night so that a neurologist could see what was going on. To round the picture out, my heartbeat, breathing rate, limb movements, body temperature, and jaw pressure would be captured as well. Cue the electrodes, sixteen in all, attached to sites ranging from my temples to my ankles. A technician slathered each spot on my head with sticky white goo, manipulating my hair into what could be called the Contemporary Einstein. She taped a forked monitor inside my nostrils, glued oval sensors to each of my cheeks, and bound what appeared to be a glowing red clothespin to my index finger. A plastic blue box heavy with wires running in and out of it hung around my neck. The process of putting all of this on my body took forty-five minutes. After she finished, the technician told me that she would be in a room down the hall watching me via a video camera fixed on the ceiling and pointed at the bed. "Try to sleep normally," she said as she closed the door. If she was aware of the irony, she didn't show it.

I tried to get comfortable. After a few minutes, I turned to my right side. Suddenly, a voice calling out from a pair of hidden speakers above the headboard echoed in the small room. "Sir, you cannot sleep on your side. You must stay on your back," the technician said. A blinking red light on the ceiling revealed the camera that gave me away. I lay there like a board, wondering when this would be over. That night, I dreamt that I was in a prison.

A few days later, I sat in the office of the neurologist who had ordered the study, a long, slender man whose oversized glasses made his face look too small for his body. He rustled through the more than three hundred pages of data collected from my night in the sleep lab, past charts showing my brain waves spiking and falling like a boom and bust stock market. His hands settled on the summary he was looking for. He studied it quietly for a few minutes. Finally, he spoke.

"Well, you certainly kick a lot."

I waited, hoping there was more to come from a test that set my health insurance company back a couple-thousand dollars.

"But beyond that, I'm not sure what we can do for you," he continued. "Your breathing is normal, so you don't have sleep apnea. You're not having seizures in your sleep. You awaken easily, that's clear, but that's not really a medical problem. I could give you a sleeping pill, but frankly I'm not sure that it's going to help."

"Do I have restless leg syndrome?" I asked, suddenly feeling like an actor in one of those commercials that tell you to ask your doctor whether a medication is right for you.

"Do your legs feel uncomfortable if you don't kick them?"

"Not really," I replied.

"Then it's not restless leg. It may be a mild case of periodic limb movement disorder, but there's not much we can do for that."

I liked the sound of the word *mild*. "So what should I do?" I asked.

"I'm going to be honest with you. There's a lot that we know about sleep, but there's a lot we don't know. If the sleepwalking continues, let's try some sedatives. But I don't want you to start taking drugs that you don't need. Try to cut down on your stress and see what happens."

I left the appointment with the vague feeling that I had been tricked. I expected science to have as thorough an understanding of sleep as it does of digestion, or any other bodily function that we can't live without. Instead I heard a doctor's disconcerting admission that he didn't know what was going on or how to stop it. It was as if my body had sleepwalked itself past the frontier.

Sleep wasn't something that we were supposed to worry about in the first years of the twenty-first century. There were bigger issues requiring attention. Technology was making the world smaller by the day, the global economy blurred the lines between one day and the next, and daily life was filled with questions over what was considered normal. Many people never gave sleep much thought, and if they did, considered it nothing more than an elegant on/off switch that the body flips when it needs to take a break from its overscheduled life. Sure, we would probably like to get more of it, and yes, we may have had a weird dream or two lately, but beyond that, the importance of sleep likely hovers somewhere near that of flossing in most of our lives: something we are supposed to do more often but don't.

Most of us will spend a full third of our lives asleep, and yet we don't have the faintest idea of what it does for our bodies and our brains. Research labs offer surprisingly few answers. Sleep is one of the dirty little secrets of science. My neurologist wasn't kidding when he said there was a lot that we don't know about sleep, starting with the most obvious question of all—why we, and every other animal, need to sleep in the first place.

Consider, for a moment, how absurd the whole idea of falling asleep is in a world of finite resources where living things resort to eating each other to survive. A sleeping animal must lie still for long stretches at a time, all but inviting predators to make it dinner (and not in a good way). Yet whatever sleep does is so important that evolution goes out of its way to make it possible. A dolphin, for instance, will sleep with half of its brain awake at a time, giving it the ability to surface for air and be on the lookout for predators while the other half is presumably dreaming. Birds, too, have adapted the ability to decide whether to put half of their brain to sleep or the whole thing. Imagine a flock of ducks sleeping at the edge of a lake. The birds at the periphery of the group will likely be sleeping with one-half of their brain awake and aware of their surroundings, keeping watch while their companions in the middle zonk out completely.

You would think, then, that sleep is a luxury that increases as you move up the food chain, and that sharper claws would equal longer dreams. But no. Lions and gerbils sleep about thirteen hours a day. Tigers and squirrels nod off for about fifteen hours. At the other end of the spectrum, elephants typically sleep three and a half hours at a time, which seems lavish compared to the hour and a half of shut-eye that the average giraffe gets each night.

The need to sleep interferes with other more biologically pressing needs, such as procreating, finding and gathering food, building shelter, and anything else you might do to ensure that your genetic line lives on. Sleep is so important, yet so poorly understood, that it led one biologist to say, "If sleep doesn't serve an absolutely vital function, it is the greatest mistake evolution ever made." That function is still a mystery. It would be nice to say that sleep is nothing more than the time when a body rests, but that wouldn't be quite right either. You can relax in a hammock on a beach all day long if you want to, but after about twenty hours you will be in pretty bad shape if you don't fall asleep and stay that way for a while. Humans need roughly one hour of sleep for every two hours they are awake, and the body innately knows when this ratio becomes out of whack. Each hour of missed sleep one night will result in deeper sleep the next, until the body's sleep debt is wiped clean.

The only thing stranger than the need to sleep is what happens when it is ignored. In 1965, a San Diego high school student named Randy Gardner stayed awake continuously for 264 hours, an eleven-day feat documented by a team of researchers from Stanford University who happened to read about his attempt beforehand in the local newspaper. For the first day or so, Gardner was able to remain awake without any prompting. But things went south quickly. He soon lost the ability to add simple numbers in his head. He then became increasingly paranoid, asking those who had promised to help him stay up why they were treating him so badly. When he finally went to bed, he slept for nearly fifteen hours straight. And yet a few weeks later, he was as good as new. To this day, he continues to be a minor celebrity in Japan.

Gardner experienced a happier ending than most subjects of sleep deprivation experiments. In the 1980s, researchers at the University of Chicago decided to find out what happens when an animal is deprived of sleep for a long period of time. In but one of the many odd tests you will find in the history of sleep research, these scientists forced rats to stay awake by placing them on a tiny platform suspended over cold water. The platform was balanced so that it would remain level only if a rat kept moving. If a rat fell asleep, it would tumble into the water and be forced to swim back to safety (or drown, an option that the researchers seemed strangely blasé about).

Fast-forward to two weeks later. All of the rats were dead. This confused the researchers, though they had a few hints that something bad was going to happen. As the rats went longer and longer without sleep, their bodies began to self-destruct. They developed strange spots and festering sores that didn't heal, their fur started to fall out in large clumps, and they lost weight no matter how much food they ate. So the researchers decided to perform autopsies, and lo and behold they found nothing wrong with the animals' organs that would lead them to failing so suddenly. This mystery gnawed at scientists so much that twenty years later, another team decided to do the exact same experiment, but with better instruments. This time, they thought, they will find out what happens inside of a rat's body during sleep deprivation that ultimately leads to its death. Again the rats stayed awake for more than two weeks, and again they died after developing gnarly sores. But just like their peers in Chicago years earlier, the research team could find no clear reason why the rats were keeling over. The lack of sleep itself looked to be the killer. The best guess was that staying awake for so long drained the animal's system and made it lose the ability to regulate its body temperature.

Humans who are kept awake for too long start to show some of the same signs as those hapless rats. For obvious reasons, no one has conducted any scientific research into whether it is possible for a person to die from extreme sleep deprivation. The closest we have come are short-term sleep deprivation studies conducted by the government, with subjects participating voluntarily or not. CIA interrogators at Guantánamo Bay, for instance, subjected dozens of enemy combatants to sleep deprivation by chaining them together and forcing them to stand for more than a day at a time. Justice Department officials later wrote in a memo that "surprisingly, little seemed to go wrong with the subjects physically."

Signs that the lack of sleep was affecting their bodies were most likely there but not apparent to the naked eye. Within the first twenty-four hours of sleep deprivation, the blood pressure starts to increase. Not long afterward, the metabolism levels go haywire, giving a person an uncontrollable craving for carbohydrates. The body temperature drops and the immune system gets weaker. If this goes on for too long, there is a good chance that the mind will turn against itself, making a person experience visions and hear phantom sounds akin to a bad acid trip. At the same time, the ability to make simple decisions or recall obvious facts drops off severely. It is a bizarre downward spiral that is all the more peculiar because it can be stopped completely, and all of its effects will vanish, simply by sleeping for a couple of hours.

I know all of this only because I walked out of that neurologist's office with more questions than answers. As I headed home, wondering if I would sleepwalk again and how badly it would hurt if I ran into something the next time, my confusion gave way to a plan. If my doctor couldn't tell me more about sleep, I reasoned, then I would go out and search for the solutions myself. A third of my life was passing by, unexamined and unaccounted for, and yet it was shrouded in mystery.

So began my adventures in the strange science of sleep. I set out to discover everything I could about a period of time that we can only conceive of as an abstraction, a bodily state that we know about but never really experience because, well, we are asleep. Once I started really thinking about sleep for the first time, the questions came in waves. Do men sleep differently than women? Why do we dream? Why is getting children to fall asleep one of the hardest parts of becoming a new parent, and is it this hard for everyone around the world? How come some people snore and others don't? And what makes my body start sleepwalking, and why can't I tell it to stop? Asking friends and family about sleep elicited a long string of "I don't knows," followed by looks of consternation, like the expressions you see on students who don't know the answers to a pop quiz. Sleep, the universal element of our lives, was the great unknown. And frankly, that makes no sense.

Despite taking up so much of life, sleep is one of the youngest fields of science. Until the middle of the twentieth century, scientists thought that sleep was an unchanging condition during which time the brain was quiet. The discovery of rapid eye movements in the 1950s upended that. Researchers then realized that sleep is made up of five distinct stages that the body cycles through over roughly ninety-minute periods. The first is so light that if you wake up from it, you might not realize that you have been sleeping. The second is marked by the appearance of sleep-specific brain waves that last only a few seconds at a time. If you reach this point in the cycle, you will know you have been sleeping when you wake up. This stage marks the last stop before your brain takes a long ride away from consciousness. Stages three and four are considered deep sleep. In three, the brain sends out long, rhythmic bursts called delta waves. Stage four is known as slow-wave sleep for the speed of its accompanying brain waves. The deepest form of sleep, this is the farthest that your brain travels from conscious thought. If you are woken up while in stage four, you will be disoriented, unable to answer basic questions, and want nothing more than to go back to sleep, a condition that researchers call sleep drunkenness. The final stage is REM sleep, so named because of the rapid movements of your eyes dancing against your eyelids. In this type of sleep, the brain is as active as it is when it is awake. This is when most dreams occur.

Your body prepares for REM sleep by sending out hormones to effectively paralyze itself so that your arms and legs don't act out the storyline you are creating in your head. This attempt at self-protection doesn't always work perfectly, and when that happens, what follows is far from pleasant. Sometimes, it is the brain that doesn't get the message. This can lead to waking up in the middle of the night with the frightening sensation that you can't move your limbs. In the Middle Ages, this was thought to be a sign that a demon called an incubus was perched on the chest. Instead, this condition is simply a flaw in the sleep cycle, a wrong-footed step in the choreography of the brain's functions that allows a person to become conscious when the body thinks the brain is still dreaming. At other times, the body doesn't fully paralyze itself like it is supposed to. This is the root of a series of problems called parasomnias, of which sleep-walking like mine is by far the most mild. Patients with REM sleep disorder, for instance, sometimes jump out of a window or tackle their nightstand while they are acting out a dream. Some patients I spoke with who have this disorder have resorted to literally tying themselves to the bedpost each night out of the fear that they will accidentally commit suicide.

Before the discovery of rapid eye movements, our understanding of sleep hadn't undergone any dramatic revisions in more than two thousand years. The Ancient Greeks believed that someone fell asleep when the brain became filled with blood, and then woke up once it drained back out again. Beyond that, they found the whole experience kind of spooky. Sleep was considered the closest a living being could come to death and still be around to talk about it afterward. The immortal family tree made this clear: Hypnos, the Greek god of sleep, was the twin brother of Thanatos, the god of death, and their mother was the goddess of night. It was probably best not to think about this too much while lying in a room on a dark and lonesome evening. Two-dozen centuries later, doctors put forth the theory that blood flowing through the head put pressure on the brain and induced sleep, a concept Plato would have readily agreed with. Philosophers in the nineteenth century introduced the novel idea that a person fell asleep when the brain ceased to be filled with stimulating thoughts or ambitions. The supposed link between sleep and an empty head fostered a suspicion of anyone who slept too much or seemed to enjoy it. In certain high-pressure jobs today, admitting that you sleep for more than five or six hours each night still looks to be a sure sign that you are not a serious person.

Whether any of us has a sleep problem or not, it is clear that we are living in an age when sleep is more comfortable than ever and yet more elusive. Even the worst dorm-room mattress in America is luxurious compared to sleeping arrangements that were common not that long ago. During the Victorian era, for instance, laborers living in workhouses slept sitting on benches, with their arms dangling over a taut rope in front of them. They paid for this privilege, implying that it was better than the alternatives. Families up to the time of the Industrial Revolution engaged in the nightly ritual of checking for rats and mites burrowing in the one shared bedroom. Modernity brought about a drastic improvement in living standards, but with it came electric lights, television, and other kinds of entertainment that have thrown our sleep patterns into chaos.

Work has morphed into a twenty-four-hour fact of life, bringing its own set of standards and expectations when it comes to sleep. As the Wall Street banker who follows investments simultaneously in Dubai, Tokyo, and London knows, if you aren't keeping up, you risk being left behind. Sleep is ingrained in our cultural ethos as something that can be put off, dosed with coffee, or ignored. And yet maintaining a healthy sleep schedule is now thought of as one of the best forms of preventative medicine.

Stanford University, one of the world's premier centers of sleep research, established the first university laboratory center devoted to treating sleep disorders in 1970. The opening of Stanford's clinic started a revolution in the way the medical field approached sleep. Until then, most doctors thought that their responsibility ended once a patient nodded off each night. By 2011 there were over seventy-five recognized sleep disorders, and the number continues to grow. Some, like sleep apnea, are so common that if they aren't present in your bedroom, there is a very good chance you will find them next door. Others are simply baffling. One rare type of prion disease called fatal familiar insomnia strikes after a person reaches the age of forty. This genetic disease has been found in only a handful of families around the world. Its chief symptom is the gradual inability to fall asleep. Within a year of the first signs of the condition, patients typically die after suffering through months of agony, beset by chronic migraines and exhaustion. Their minds remain clear and unaffected until death.

There is more to sleep than medical curiosities, however. This is a book about the largest overlooked part of your life and how it affects you even if you don't have a sleep problem that sends you into a wall in the middle of the night. I began my research into sleep with the self-serving intention of finding a way to prevent future run-ins. But as I spent more time investigating the science of sleep, I began to understand that these strange hours of the night underpin nearly every moment of our lives. Police officers, truck drivers, and emergency-room physicians, for instance, are turning to sleep researchers for help in navigating sleep's effects on the brain's decision-making process. If you have ever flown on an airplane, gone to a hospital, or driven on a highway at night—or plan on doing so in the future—then you have a vested interest in how companies and organizations try to prevent costly and deadly accidents caused by something as manageable as fatigue. School districts across the country, meanwhile, have changed the time that the first bell rings in the wake of research showing that simply starting the school day later leads to significantly higher SAT scores. And new studies suggest that learning a new skill or finding a solution to a problem may simply be an outcome of the time that we spend dreaming each night.

Because of the number of new findings in such a short time span, today's researchers believe that they are in the golden age of their field. Sleep is now understood as a complex process that affects everything from the legal system to how babies are raised to how a soldier returning from war recovers from trauma. And it is also seen as a vital part of happiness. Whether you realize it or not, how you slept last night probably has a bigger impact on your life than what you decide to eat, how much money you make, or where you live. All of those things that add up to what you consider you—your creativity, emotions, health, and ability to quickly learn a new skill or devise a solution to a problem—can be seen as little more than by-products of what happens inside your brain while your head is on a pillow each night. It is part of a world that all of us enter and yet barely understand.

Sleep may not immediately come across as the most adventurous topic to investigate. After all, people who are sleeping are usually just lying there, making it very hard to interview them. What could possibly be interesting about that? My aim is to convince you otherwise by taking you on a tour of often odd, sometimes disturbing, and always fascinating things that go on in the strange world of sleep, a land where science is still in its infancy and cultural attitudes are constantly changing. I will take you through the story of a night, starting with the unrecognized forces at work in your bedroom as you fall asleep and ending with the latest research into what goes into a good night's rest.

This is not your typical advice book filled with ten easy steps to perfect sleep. But you will come away with a new understanding of all that goes on in your body while you are sleeping and what happens when you neglect sleep for too long. Hopefully, this will inform your future decisions affecting everything from your health to your wallet. You don't have to take my word for it. By the end of this book, you will have met, among others, dream researchers, professional sports trainers, marriage counselors, pediatricians, constitutional scholars, gamblers, and a university professor who investigates what could only be called sleep crime.

I never found the cure for sleepwalking that I was looking for, though I did learn what I could do to make it less likely to happen again without resorting to medication. But no matter what steps I take, or how much yoga I do to relax myself before bedtime, I very well may wake up once again in the middle of the night, disoriented and away from my bed. On the other hand, I may never sleepwalk again. That's the bizarre beauty of sleep, a seemingly simple part of life with more possible outcomes each night than you can imagine. I've been to military bases and corporate headquarters, campus labs and convention centers, all in search of what we can learn from this curious and universal fact of life if we took the time to examine it.

Sleep isn't a break from our lives. It's the missing third of the puzzle of what it means to be living.

2

Light My Fire

If you wanted to find Roger Ekirch for most of the 1980s and 1990s, the first and best place to look was between the gray stone walls of the Virginia Tech University library. A young professor of history who taught courses about life in the early United States, Ekirch spent most of his days giving lectures to undergraduates about the early slave trade or the once-booming pirate economy of the Atlantic. But whenever he could, he sequestered himself among the rare-book collection. It was there that he could indulge a topic that had intrigued him since graduate school: the history of the night.

At the time, most historians would have readily agreed that human activity after the sun went down was reduced to "no occupation but sleep, feed and fart," as Thomas Middleton, a playwright who was friends with Shakespeare, once so eloquently put it. But Ekirch continued the lonely work of prying open the pages of mildewing books, noting any hints that something interesting happened after the close of each day. He didn't know he was on the path of a major breakthrough that would change our conception of how the human brain is built for sleep. He was a history professor, after all, whose only understanding of sleep consisted of knowing that he liked it. But as he searched through plays, wills, and all of the other assorted artifacts of daily life that had accumulated over the last thousand years in Europe, he realized that the sun's fall into the horizon set the stage for a bizarre twelve hours.

Nightfall on an average day of the week brought about a fear so harrowing to a villager in medieval Europe that we can scarcely conceive of it today. At the first hints of sunset, farmers raced to get inside a city's walls before they were locked at night. Anyone not fast enough would have to survive the dark hours in the wilderness alone, fending off robbers, wolves, and the ghosts and devils lurking around every corner.

The cities weren't much safer. If you were to find yourself on the streets at night, the logical assumption was that everyone you encountered was intent on robbing or killing you. Striking first became the best option. Past nightfall, "clashes of all sorts became likely when tempers were shortest, fears greatest, and eyesight weakest," Ekirch noted. He found stories of servants stabbing each other in the armpit "without provocation," merchants getting into sword fights with their neighbors on the streets of London, and the sound of dead bodies splashing into the canals of Venice—all a common part of life after dark. In these times, when most everyone who ventured outside at night did so armed with at least a knife, a polite greeting was less of a formality and more of a way to stay alive.

The hours of the night were so starkly different that they had their own cultural rhythms. Townspeople who took pride in their ability to fend for themselves during the day willfully submitted to curfews, literally locking themselves into their homes at night. Rural farmers who would never see an ocean in their lifetimes knew how to tell time and direction from the stars, just like sailors. Monarchs and bishops demonstrated their authority over the elements by staging elaborate ceremonies and dances illuminated by hundreds of torches, dazzling the eyes of peasants who relied on stinky, smoky, and dim candles to light their small houses.

Yet something puzzled Ekirch as he leafed through parchments ranging from property records to primers on how to spot a ghost. He kept noticing strange references to sleep. In the *Canterbury Tales*, for instance, one of the characters in "The Squire's Tale" wakes up in the early morning following her "first sleep" and then goes back to bed. A fifteenth-century medical book, meanwhile, advised readers to spend the "first sleep" on the right side and after that to lie on their left. And a scholar in England wrote that the time between the "first sleep" and the "second sleep" was the best time for serious study. Mentions of these two separate types of sleep came one after another, until Ekirch could no longer brush them aside as a curiosity. Sleep, he pieced together, wasn't always the one long block that we consider it today.

From his cocoon of books in Virginia, Ekirch somehow rediscovered a fact of life that was once as common as eating breakfast. Every night, people fell asleep not long after the sun went down and stayed that way until sometime after midnight. This was the first sleep that kept popping up in the old tales. Once a person woke up, he or she would stay that way for an hour or so before going back to sleep until morning—the so-called second sleep. The time between the two bouts of sleep was a natural and expected part of the night and, depending on your needs, was spent praying, reading, contemplating your dreams, urinating, or having sex. The last one was perhaps the most popular. One sixteenth-century French physician concluded that laborers were able to conceive several children because they waited until after the first sleep, when their energy was replenished, to make love. Their wives liked it more, too, he said. The first sleep let men "do it better" and women "have more enjoyment."

Ekirch was faced with the classic crisis of the scholar: here in front of him was mounting evidence that how we sleep today is nothing like the sleep of our ancestors. Yet saying that the whole of the industrialized world sleeps unnaturally was a big leap, especially for a professor who was more versed in the agrarian economy of the American colonies than in neuroscience. Even years later, Ekirch couldn't be sure that he would have publicized his findings without a bit of luck. "I would have hoped that I would have had enough confidence in my research to go ahead with the idea on my own," he told me, sounding like a man trying to build his confidence through a barrage of words.

Fortunately for him, he didn't have to. About three hundred miles away, a psychiatrist was noticing something odd in a research experiment. Thomas Wehr, who worked for the National Institute of Mental Health in Bethesda, Maryland, was struck by the idea that the ubiquitous artificial light we see every day could have some unknown effect on our sleep habits. On a whim, he deprived subjects of artificial light for up to fourteen hours a day in hopes of re-creating the lighting conditions common to early humans. Without lightbulbs, televisions, or street lamps, the subjects in his study initially did little more at night than sleep. They spent the first few weeks of the experiment like kids in a candy store, making up for all of the lost sleep that had accumulated from staying out late at night or showing up at work early in the morning. After a few weeks, the subjects were better rested than perhaps at any other time in their lives.

That was when the experiment took a strange turn. Soon, the subjects began to stir a little after midnight, lie awake in bed for an hour or so, and then fall back asleep again. It was the same sort of segmented sleep that Ekirch found in the historical records. While sequestered from artificial light, subjects were shedding the sleep habits they had formed over a lifetime. It was as if their bodies were exercising a muscle they never knew they had. The experiment revealed the innate wiring in the brain, unearthed only after the body was sheltered from modern life. Not long after Wehr published a paper about the study, Ekirch contacted him and revealed his own research findings.

Wehr soon decided to investigate further. Once again, he blocked subjects from exposure to artificial light. This time, however, he drew some of their blood during the night to see whether there was anything more to the period between the first and second sleep than an opportunity for feudal peasants to have good sex. The results showed that the hour humans once spent awake in the middle of the night was probably the most relaxing block of time their lives. Chemically, the body was in a state equivalent to what you might feel after spending a day at a spa. During the time between the two sleeps, the subjects' brains pumped out higher levels of prolactin, a hormone that helps reduce stress and is responsible for the relaxed feeling after an orgasm. High levels of prolactin are also found in chickens while they lay atop their eggs in a contented haze. The subjects in Wehr's study described the time between the two halves of sleep as close to a period of meditation.

Numerous other studies have shown that splitting sleep into two roughly equal halves is something that our bodies will do if we give them a chance. In places of the world where there isn't artificial light—and all the things that go with it, like computers, movies, and bad reality TV shows—people still sleep this way. In the mid-1960s, anthropologists studying the Tiv culture in central Nigeria found that group members not only practiced segmented sleep, but also used roughly the same terms of first sleep and second sleep.

You would think that investigations showing that our modern sleep habits run contrary to our natural wiring would be a pretty big deal. But almost two decades after Wehr's study was published in a medical journal, many sleep researchers—not to mention your average physician—have never heard of it. When patients complain about waking up at roughly the same time in the middle of the night, many physicians will reach for a pen and write a prescription for a sleeping pill, not realizing that they are medicating a condition that was considered normal for thousands of years. Patients, meanwhile, see waking up as a sign that something is wrong. Without knowing that sleep is naturally split into two periods, it's hard to blame them.

Why do roughly six billion humans sleep in a way that is contrary to what worked for millions of years? Because of a product that was once revolutionary and now costs less than two dollars: the lightbulb. The lamp next to your bed contains a device that has changed human sleep perhaps forever, and ushered in a new world of health problems that come from an overabundance of light. Nearly every aspect of modern life originated in a complex of weathered brick buildings surrounded by a black metal fence in northern New Jersey. Here, in an idea factory that predated the startups of Silicon Valley, an inventor with a talent for self-promotion named Thomas Alva Edison forged the devices that upended how our bodies are designed to sleep.

Of course, some artificial lights were in use before Edison came around. In 1736, the city of London took a giant leap forward by installing five thousand gaslights in its streets, taming the city's long-held fear of the dark and allowing shopkeepers to stay open past ten at night for the first time. Other cities followed as gaslights became a mark of cosmopolitan prestige. By the beginning of the Civil War, there were so many gas lamps on the streets of New York City that it was as common to venture into the night as it was during the daytime. Theaters, operas, and saloons lit by gaslights stayed open until the early morning as the newly lit streets promised a safe ride home. Homes, too, glowed from the light of flames.

Yet all of the artificial light in use around the world before Edison developed his lightbulb amounted to the brightness of a match compared to the lights of Times Square. Edison's career as an inventor began when, as a bored teenage telegraph operator, he tried to come up with ways to send more than one message at a time on the machine. A few years later he made a name for himself by inventing the phonograph. In the first instance of what would become a defining trend of his life, Edison didn't quite realize the popular appeal of the technical wonder he created. He saw the phonograph as a way for busy executives to dictate letters that would then be listened to and transcribed by aides. The invention became a commercially viable product only after dealers set up arcades where customers could listen to recorded music for five cents apiece. Edison had no idea that he had just unleashed the genesis of America's mass entertainment industry, in part because he couldn't partake in it: a hearing loss sapped his enjoyment of music.

Around this same time, French inventors installed what was known as arc light—so called because it sent currents on an arc across a gap—on the streets of Lyon. The light wasn't anything you would want in your kitchen, unless you had a desire to burn the house down. Arc light was a barely controlled ball of current, closer to the intense, white light from a welder's torch than the soft glow of the bulb in your refrigerator. The contraption generated plenty of light, but it wasn't pretty. In Indiana, four arc lights installed on top of a city's courthouse were said to be bright enough to illuminate cows five miles away. The town of San Jose, California, built a twenty-story tower and put arc lights on it. Confused birds crashed into it and eventually made their way to the tables of the city's restaurants.

Armed with a little fame and money from the phonograph, Edison set off to invent a better form of artificial light than the arc lamp. His goal was to domesticate light, making it simple enough that a child could operate it and safe enough that accidentally leaving it on all night wouldn't start a fire. He designed a lightbulb that glowed from electric currents passing through a horseshoe-shaped wire set in a vacuum, which essentially kept it from melting or catching on fire. His technique wasn't necessarily the smartest or the best of the approaches to lightbulbs at the time, but he knew how to sell himself as part of the product. He slyly cultivated a public image as a wizard of inventions by handing out ownership stakes in his companies to reporters who made the trek out to see him at his lab in Menlo Park, New Jersey, and later wrote flattering articles. Edison made sure that everyone had an ample chance to hear his last name by inserting it into the companies he founded to back each new project. One of them, the Edison Electric Co., eventually morphed into General Electric.

Edison's light became the standard of the world because it was cheap, safe, and just powerful enough to be comfortable. Unlike arc light, the lightbulb's beauty was its small capacity. It wasn't bright enough to reach cows a few miles away, but it had an even, steady glow that could illuminate a living room full of guests. Within a few years of its invention, a parade of men walked down the streets of New York wearing bulbs on their heads to demonstrate that light no longer had to come from flames.

If all that Edison did was perfect artificial light, he would have undoubtedly changed the course of sleep history. But he didn't stop there. Not quite satisfied with remaking how we experience night, Edison also had a singular role in revolutionizing entertainment. He perfected the phonograph and later developed one of the first motion-picture cameras. Through these inventions, Edison created an utterly new experience: watching or listening to a person who wasn't live in front of you. Paying customers could now see the performances of boxers, singers, and orchestras that were recorded, creating a democratic world of celebrity where everyone with a nickel could see or hear world-class entertainment. The best performers in the world were taken out of exhibition halls and available in the living room.

Thanks to Edison, sunset no longer meant the end of your social life; instead, it marked the beginning of it. Night shook off its last associations with fear and became the time when all the good stuff happens. Life could function just as well at eleven o'clock at night as it did at eleven in the morning, with darkness no longer getting in the way. The world responded to these extra hours of possibility by acting like college students spending their first month in a dorm. Sleep took a backseat to nightlife and other more important priorities, and it has never regained its former place. Manufacturers, too, recognized that they could double production without sacrificing quality by running shifts overnight while lightbulbs provided illumination. Within twenty years of their development in Edison's laboratory, lightbulbs were hanging from the ceilings of assembly lines where some of the first graveyard-shift workers tried to stay awake on the job. There was no longer a need to leave the workbench idle just because the sun went down. The twenty-four-hour workforce was born.

Edison saw no problem as he watched the natural rhythms of sleep irrevocably change. For a reason that was never quite clear, he thought that sleep was bad for you. "The person who sleeps eight or ten hours a night is never fully asleep and never fully awake," he wrote. "He has only different degrees of doze through the 24 hours." Extra sleep—defined as anything more than the three or four hours that Edison claimed he slept each night—made a person "unhealthy and inefficient." Edison saw his lightbulb as a form of nurture and believed that all one had to do was "put an undeveloped human being into an environment where there is artificial light and he will improve."

Life, in his eyes, was like an assembly line where any downtime could be only wasteful. Not that Edison required less sleep than the rest of us. He napped throughout the day and night, sometimes falling asleep on a workbench in his laboratory and then claiming the next day that he had worked through the night. Visitors to his lab in Menlo Park can still see his small cot and pillow tucked away in a corner.

Combined with his lightbulb, Edison's idea that sleep was a sign of laziness refashioned the way the world worked. Some of the earliest battles in the labor movement in the United States were over how long a night shift could last. Places that clung to their traditional sleeping schedules were quickly derided as backwaters filled with people who weren't fit for the industrialized world.

Now, about a hundred years later, we have so much artificial light that after a 1994 earthquake knocked out the power, some residents of Los Angeles called the police to report a strange "giant, silvery cloud" in the sky above them. It was the Milky Way. They had never seen it before, and with good reason: LA is lit up at night by so many streetlights, billboards, hotels, cars, sports stadiums, parking lots, and car dealerships that airplanes can see the glow of the city from two hundred miles away. Angelenos aren't alone. Two-thirds of the population of the United States and half of Europe live in areas where the night sky shines too brightly to see the Milky Way with the naked eye. In the United States, ninety-nine out of every hundred people live in an area that meets the standard of light pollution, which is what astronomers call it when artificial lights make the night sky more than ten times brighter than it would be naturally.

If all lights did was to make it easier to find things at night, there wouldn't be much to get worked up about. But the sudden introduction of bright nights during hours when it should be dark threw a wrench into a finely choreographed system of life. Some ten thousand confused birds—which, like moths, are attracted to bright lights—die each year after slamming into glowing skyscrapers in Manhattan. More than one hundred million birds crash into brightly lit buildings every night across North America. Biologists now point to artificial light as a threat to the living environments of organisms as varied as sea turtles, frogs, and trees.

Let's not kid ourselves: the animal that you are most concerned with is the one reading this book. Just like every other living being, you too are affected by the glow of streetlamps and skyscrapers. Electric light at night disrupts your circadian clock, the name given to the natural rhythms that the human body developed over time. When you see enough bright light at night, your brain interprets this as sunlight because it doesn't know any better. The lux scale, a measure of the brightness of light, illustrates this point. One lux is equal to the light from a candle ten feet away. A standard 100-watt lightbulb shines at 190 lux, while the lighting in an average office building is 300 lux. The body's clock can be reset by any lights stronger than 180 lux, meaning that the hours you spend in your office directly impact your body's ability to fall asleep later. That's because your body reacts to bright light the same way it does to sunshine, sending out signals to try to keep itself awake and delay the nightly maintenance of cleanup and rebuilding of cells that it does while you are asleep. Too much artificial light can stop the body from releasing melatonin, a hormone that helps regulate sleep.

Poor sleep is just one symptom of an unwound body clock. Circadian rhythms—which you will learn much more about in a later chapter—are thought to control as many as 15 percent of our genes. When those genes don't function as they should because of the by-products of artificial light, the effects are a rogue's gallery of health disorders. Studies have linked depression, cardiovascular disease, diabetes, obesity, and even cancer to overexposure to light at night. Researchers know this, in part, from studying nurses who have spent years working the graveyard shift. One study of 120,000 nurses found that those who worked night shifts were the most likely to develop breast cancer. Another found that nurses who worked at least three night shifts a month for fifteen years had a 35 percent greater chance of developing colon cancer. The increased disease rates could not be explained as a by-product of working in a hospital.

In one of the most intriguing studies, researchers in Israel used satellite photos to chart the level of electric light at night in 147 communities. Then, they placed the satellite photos over maps that showed the distribution of breast cancer cases. Even after controlling for population density, affluence, and other factors that can influence health, there was a significant correlation between exposure to artificial light at night and the number of women who developed the disease. If a woman lived in a place where it was bright enough outside to read a book at midnight, she had a 73 percent higher risk of developing breast cancer than a peer who lived in a neighborhood that remained dark after the sun went down. Researchers think that the increased risk is a result of lower levels of melatonin, which may affect the body's production of estrogen.

There could be more discoveries on the horizon that show

detrimental health effects caused by artificial light. Researchers are interested in how lights have made us less connected to the changing of the seasons. "We've deseasonalized ourselves," Wehr, the sleep researcher, said. "We are living in an experiment that is finding out what happens if you expose humans to constant summer day lengths."

The long glow of artificial lights and the short shrift given to sleep are now dominant parts of the global economy, forcing cultures that have long cherished a midday nap to conform to a world of work that Edison would approve of. Though midday naps are most closely linked with Spain and other Latin cultures, they were once popular throughout Europe, Africa, and Asia. Even today, most state-owned firms in China give their workers two hours for lunch. The first is used for eating and the second, for sleeping. One persistent gripe among managers of multinational corporations growing quickly in that country is that their employees put their heads down on their desk after lunch and sleep for thirty minutes or so.

And yet economics may eventually catch up to China as they have to the siesta in Spain. There, the tradition of taking a midday nap was curtailed in 2006 when the federal government reduced the customary three-hour lunch break for government employees to one hour in hopes that private businesses would follow. The idea was to keep Spaniards at their desks at the same time that the rest of Europe was in the office. Though some areas still largely close down at siesta time, what was once a hallmark of Spain's culture has in some ways been reduced to a tourist ploy. In 2010, for instance, a shopping center in Madrid set up a bunch of blue couches and held what it called the Siesta National Championship. Anyone walking by was free to change into blue pajamas and take a nap. Contestants were rated on how long they slept and how loudly they snored. The idea was to show potential visitors that Spain was a place where it was so relaxing that anyone could fall asleep in an instant. But, coming in the middle of a financial crisis, the scheme didn't go over so well. One British visitor fumed to the local paper: "We're talking about the potential of a collapsing euro. We're talking about surging debt, and people are still wanting to preserve the tradition of sleeping while the rest of the world is working?"

It was a fair point, but the idea of working without paying attention to the need for sleep results in its own form of failure. Hospitals, which should know better, are among the worst culprits. In the first part of the 2000s, professors from Harvard Medical School and Brigham and Women's Hospital in Boston rounded up nearly twenty thousand doctors who were in their first year of residency and asked them to fill out a simple survey about their work lives. Work was pretty much all these interns did. Many had shifts that lasted thirty straight hours. Spending a hundred hours a week on the clock wasn't unheard of. These doctors were no doubt trained professionals at the hospital, capable of performing their jobs under stress.

But once they got on the road to go home, it was a different story. The study found that interns who worked more than twenty-four straight hours were twice as likely to get in a car accident than a colleague who worked a shorter shift. The higher number of long shifts the doctors worked, the more likely they were to become a danger on the roadway. Interns who worked at least five long shifts a month were twice as likely to fall asleep while driving a moving vehicle, and three times more likely to fall asleep while stopped at a red light, than a colleague who worked fewer hours.

Employers who want or need to keep their businesses open at all times are realizing that they are have to deal with the equivalent of sleepy doctors causing accidents if they continue to expect employees to work extra-long shifts regularly. That is where Martin Moore-Ede comes in. A former professor at Harvard Medical School, Moore-Ede now runs one of the largest companies in the growing field of fatigue management. More than half of the companies in the Fortune 500, and a Super Bowl–winning team, have asked Moore-Ede's company, Circadian, to develop working environments for their businesses that allow a worker's body to function at high levels despite the demands of sleep and exposure to artificial light.

He spoke with me while in his office in Cambridge, Massachusetts. With glasses perched on his nose and a receding hairline that hints at his age, Moore-Ede looked very much like the former professor that he is. The last year had been very good for him. His company had expanded, and had offices in Australia, Japan, the United Kingdom, the Netherlands, and Germany. His client list included Exxon-Mobil, Chevron, and American Airlines. The blue-chip companies of the world were paying him sums of money that he would only call "not inexpensive" to train their multinational workforces. More business was coming, thanks to government regulations in the United States and the United Kingdom that took effect in 2010 and required businesses in certain fields to have a fatigue management policy in place. Similar rules were already in place in Australia, Canada, and parts of Europe.

Solving the problem of sleep-deprived employees entails a lot more than giving a tired worker a pillow and a place to lie down, though that would certainly help. Fatigue management is one of those lines of work, like running a hotel, that sounds very easy until you try to do it yourself. Because of how the body's clock works and how the brain reacts to artificial light, expecting someone to sleep soundly at any time of day or night isn't always possible. The chief reason is that, unlike teenage bodies, adult bodies are not built to sleep past noon. A study by researchers in Sweden found that even in ideal sleeping conditions, subjects who would normally sleep eight hours if they went to bed at eleven o'clock at night tend to sleep only six hours if they wait until three in the morning to fall asleep. Timing trumps being tired. Even making a person exhausted beforehand doesn't change the body's awareness of the clock. In one study, subjects were kept up all night and only allowed to go to sleep at eleven in the morning. Most slept for just four hours. Though exhausted, their bodies wouldn't let them stay in dreamland.

Moore-Ede's job often boils down to challenging conceptions about the workplace that haven't been updated since Edison's time. Sometimes that leads to arguments with employers who can't accept that letting workers sleep while on the clock can be a productive use of time. "The railroad industry almost threw me out of the room when I suggested that engineers should take a brief nap rather than have to stay up continuously," he told me with obvious pride in his voice.

But more often than not, he uses numbers to speak to businesspeople in the language they understand: money. He discovered that one transportation company was paying out $32,000 in accident costs per every million miles its workers and equipment traveled. The company clocked hundreds of millions of miles a year, which made these costs far from trivial. Moore-Ede developed a staffing model that restricted long work shifts and required workers to pass awareness tests to prove that they weren't in danger of falling asleep on the job. Within months, accident costs plummeted to only $8,000 per million miles. Overall, the company's return on its investment was greater than ten to one.

Work schedules that recognize the importance of sleep and the constraints of the human body can also save lives. This was clear after an explosion occurred in Texas City, a suburb of Houston with a four-mile stretch that comprises one of the largest industrial sites in the world. Metal towers and giant vats are laid out in a long rectangle that extends to the water's edge. In early March of 2005, a visitor to the center of Texas City would have found a refinery owned and operated by BP, the British oil giant, with a capacity of processing 460,000 barrels a day—the third largest refinery of its kind in the United States. Later that month, liquid began backing up in a section of the plant that was used to manufacture highly explosive jet fuel. Three hours after the malfunction began, the level of liquid in one of the refinery's towers was at least twenty times higher than it should have been. It suddenly exploded. Fifteen workers were killed instantly. Another 170 were injured.

Investigators on the scene identified a number of reasons for the large number of fatalities, including the lack of an early-warning system and poor management policies that often overlooked posted safety rules. But Moore-Ede saw something else when he searched through work logs at the plant. The men and women on duty that day in Texas City were exhausted. Some operators were working a twelve-hour shift for the thirtieth day in a row, leaving them so sleep deprived that their brains were unable to recognize the signs that they were nearing a major catastrophe.

The explosion in Texas City was the accident that changed how the world's oil companies approach sleep. "The industry said, 'We have to get ahead of this curve or we'll get some government regulation on this issue that we're not going to want to live with,'" Moore-Ede, who served as the group's scientific advisor, told me. In 2010, the giant international oil companies agreed to install a fatigue management system at every major plant that will reduce mandatory overtime, train supervisors to recognize when an employee is close to nodding off, and give employees a chance to admit fatigue without worrying that they will lose their jobs. Moore-Ede predicts that fatigue management officers will soon be a common position in human relations departments at multinational corporations around the world. If that happens, it will be the latest in the long string of fallouts from Edison's invention.

There is little chance that we will go back to the way our bodies are meant to approach sleep. Even those who argue that recreating ancient life would solve many contemporary health problems draw the line at attempting to replicate the first and second sleep. One December day I spoke with Loren Cordain, a professor at Colorado State University. Cordain is widely acknowledged as one of the creators of what is known as the Paleo Diet. By eating like humans did before the development of agriculture, Cordain believes that we can evade health issues such as obesity, diabetes, and degenerative diseases. His diet consists of meat, seafood, and eggs, but no potatoes or grains that require cultivation. Cordain thinks that modern lifestyles are leading us to disease and discomfort, but he stops short of changing his sleep habits and reverting to a world without artificial light. "We're not hunter-gatherers anymore," he told me. "We could never duplicate that world. Nor would we want to. It's an absolutely awful experience with disease and insects and snakebites. We are people living in the Western world under Western conditions."

Of course, figuring out how to sleep in the Western world, lights or no lights, is no picnic. In the next chapter, you will meet the professor who found himself on the evening news after he said that men shouldn't sleep in the same bed as their wives. Who knew that was all it took to make a sleep scientist famous?

3

Between the Sheets

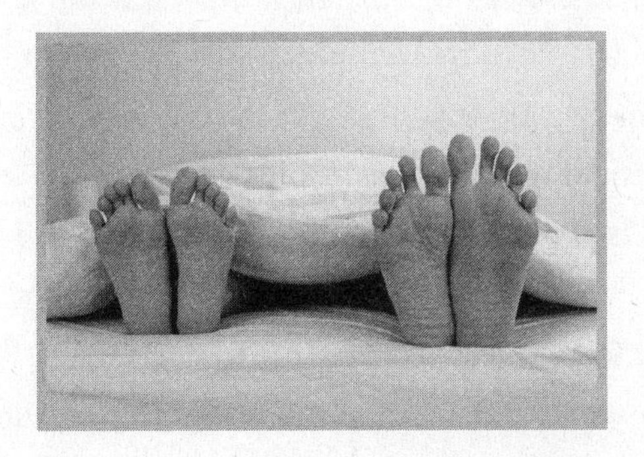

The British Science Festival is a pretty big deal in the world of European scientists. An event held annually since 1831, except during times of war, the festival's history includes the first use of the term *dinosaur*, the first demonstration of wireless transmission, and an important early debate on Darwinism. One week in late September of 2009, thousands of researchers left their labs and set off for Guildford, the town about thirty miles outside of London where the festival was held that year, to present their latest findings and to gossip about faculty openings. It wasn't the type of event—like, say, the Oscars, or the Cannes International Film Festival—that tabloid editors circle on their calendars because they expect something big to happen. Yet the minute Neil Stanley opened his mouth, the humble gathering of doctorates transformed into international news.

The kicker was the scientific suggestion that sharing a bed with someone you care about is great for sex, but not much else. Stanley, a well-regarded sleep researcher at the University of Surrey whose gray-thinning hair hinted at his more than two decades in the field, told his listeners that he didn't sleep in the same bed as his wife and that they should probably think about getting their own beds, too, if they knew what was good for them. As proof, he pointed to research he conducted with a colleague which showed that someone who shared a bed was 50 percent more likely to be disturbed during the night than a person who slept alone. "Sleep is a selfish thing to do," he said. "No one can share your sleep."

There just wasn't enough room, for one thing. "You have up to nine inches less per person in a double bed than a child has in a single bed," Stanley said, grounding his argument in the can't-argue-with-this logic of ratios. "Add to this another person who kicks, punches, snores and gets up to go to the loo and is it any wonder that we are not getting a good night's sleep?" He wasn't against sex, he assured his audience—only the most literal interpretation of sleeping together. "We all know what it's like to have a cuddle and then say, 'I'm going to sleep now,' and go to the opposite side of the bed. So why not just toddle off down the landing?"

Stanley then turned to the effects of all of those poor nights of sleep, charting a sad lineup of outcomes ranging from divorce to depression to heart disease. But there was hope, he said. Because sleep is as important as diet and exercise, maximizing our rest meant that we would be fitter, smarter, healthier—the sort of people, in short, we would want to share a cuddle with. "Isn't it much better when someone tiptoes across the corridor for a snuggle because they want to, rather than snoring, farting and kicking all through the night?" Stanley wondered.

The suggestion was eminently practical, but a social grenade. Newspapers begged him to write opinion pieces. Psychologists and marriage counselors debated on television what sleeping in separate beds said about the state of a relationship. From the response to his talk, it was clear that Stanley was not the only one who had had enough of ongoing nocturnal battles over snoring, blankets, temperature control, lighting, and every compromise that comes with lying next to another person every night. He became famous for daring to say what many had always thought: even the most lovely person in the world can turn into an enemy taking up space on a mattress once sleep is at stake.

This is far from romantic. The average person in a relationship is inclined to sleep next to his or her partner regardless of the drawbacks, a phenomenon that shows up in studies of sleep quality. In a test conducted by one of Stanley's colleagues, researchers monitored couples over several nights of sleep. Pairs were split up and sent to sleep in separate rooms for half of the test, and then allowed to come back to the shared mattress for the rest. When asked to rate their sleep quality when they woke up, subjects tended to say that they had a better night's sleep on the nights when their partner was next to them.

But their brain waves suggested otherwise. Data collected from the experiment found that subjects not only were less likely to wake up during the night but also spent almost thirty additional minutes in the deeper stages of sleep on nights when they had a room to themselves.

Here was a case where the heart seemed to conflict with the brain and the body. Despite the benefits of better-quality sleep when given their own rooms, subjects in the test consistently chose sleeping next to their partners. The question was, why? Was there something innately satisfying about sleeping next to someone else that couldn't be found on a chart of brain waves? Or was it simply habit?

The answer to that question is more complicated than it first appears, in part because of the ever-changing conceptions of what constitutes a healthy relationship. Beds, as you may not be surprised to learn, played a large part in the history of monogamy. Before the start of the Industrial Age, a mattress and its frame were often the most expensive purchases made in a lifetime, and for good reason. The common bed was where the most significant events of life happened: sex, births, illnesses, and death. The mattress—whether stuffed with feathers, straw, or sawdust—was where one came into the world and was the last stop on the ride out. Within a family, who slept on what easily corresponded to the everyday hierarchy of family life. Parents would get the most comfortable spot, often the family's only mattress, while children made do with whatever soft materials they could find. The nightly ritual of sleep meant rounding everyone up, checking the room for rats and bugs, and blowing out a candle. Few had their own rooms, but sleeping indoors rather than outside was considered a small luxury. Those who could afford otherwise were limited to the aristocracy, a class that often chose to have separate sleeping quarters for marriage partners because few unions were based on love in the first place.

This began to change in the Victorian era, a time we now recognize as the start of the modern age in which old habits were rapidly shed and reconstituted into a new way of life. In England and elsewhere, science took on a new air of professionalism, and culture put an emphasis on progress. Cities expanded, owing to the benefits of industrialization, and the emerging middle class gained the means to emphasize cleanliness and sanitation in response to the grime of urban life.

Hygiene became paramount. Science had yet to accept that germs spread disease, but demonstrations in the power of electricity and radio waves hinted at the power of an unseen world. As a result, influential public health figures believed that sickness was caused by so-called bad air, a theory called miasma. Edwin Chadwick, who was eventually knighted for his part in directing the cleanup of the sewer system while he was sanitary commissioner of London, believed until his death that the source of cholera was stench alone. "All smell is disease," he wrote.

Those theories soon filtered into the bedroom. "The home, far from being a simple haven of safety and calm to which those tossed on the turbulent seas of public life could retreat, was seen as a place of actual and potential danger," noted Hilary Hinds, a professor at Lancaster University who has studied the era. In 1880, for example, a self-proclaimed British health expert known as Dr. Richardson spent thousands of words in his influential international bestseller *Good Words* on the subject of keeping a bedroom sanitary. He advised his readers that sleeping next to someone else was a potential death trap. "At some time or other the breath of one of the sleepers must, in some degree, affect the other; the breath is heavy, disagreeable, it may be so intolerable that in waking hours, when the senses are alive to it, it would be sickening, soon after a short exposure to it," Richardson wrote. "Here in bed with the senses locked up, the disagreeable odour may not be realised, but assuredly because it is not detected it is not less injurious." Sleep, in other words, was when your partner's bad breath could strike just when your defenses were down. Richardson believed that "the system of having beds in which two persons can sleep is always, to some extent, unhealthy."

If bad air wasn't enough, there was also a brimming fear that a spouse could unwittingly steal his or her partner's invisible electrical charges. The health concerns of sleeping next to another person captivated a doctor named R. B. D. Wells, whose chief specialty was phrenology—a soon-to-be-discarded pseudoscience which held that the size of the head determined a person's intelligence and personality traits. Wells conceded that it was possible for couples to share a bed successfully, but those cases were rare. "Two healthy persons may sleep together without injury when they are of nearly equal age, but it is not well for young and old to sleep together," he wrote. "Married couples, between whom there is a natural affinity, and when one sex is of a positive and the other of a negative nature, will be benefited by the magnetism reciprocally imparted; but, unhappily, such cases of connubial compatibility are not common." Differing magnetic natures in a couple would inadvertently lead to the drainage of the "vital forces" from one partner throughout the night, a silent health threat that would leave the weakened party "fretful, peevish, fault-finding and discouraged." The clashing of electrical forces each night, over a lifetime, would be irreversible. "No two persons, no matter who they are, should habitually sleep together. One will thrive and the other will lose."

But there was a remedy—what Dr. Richardson called the "single-bed system," or what we would now recognize as a twin bed. These slender mattresses, built for one, gave a reassuring sense of distance from a spouse whose electrical charges or breath may be suspect. Each member of the couple was in a cleaner, less polluted environment, giving both an advantage in the daily battle of survival that Darwin had recently made so clear. Other experts readily joined Richardson's cause. "Such a thing even as a double bed should not exist," admonished one of his contemporaries. The public was convinced. Middle-class customers flocked to the new beds and their iron frames (wood, after all, was a building material whose hygiene was also suspect).

Dr. Richardson's solution proved so popular that even the eventual rejection of the miasma theory didn't stop the march of the twin beds. They were no longer necessary if the body's bad air alone was not the cause of disease, but they had other things going for them. For one, these beds evoked a certain modern sensibility and taste on the part of the buyer. Department stores ran ads aimed at middle-class shoppers that put twin beds squarely in the middle of chic bedrooms. The end of the sanitary craze allowed furniture stores to boast of new mattresses and frames that "combine[d] all the hygienic advantages of the metal, with the artistic possibilities of the wooden bedstead."

But no discussion of beds was ever about furniture alone. Sex was always a consideration as well. And for many who had enough money to furnish their homes with more than function in mind, their approach to sex constituted a big part of who they were. "One distinctive characteristic of the emerging middle class was its emphasis on its unique sexual morality," Stephanie Coontz, a professor of family history at Evergreen State College, told me. "They constructed their class identity on the basis of their moral rectitude, in contrast to the 'immoral' poor and the 'debauched' aristocracy. Their insistence on sexual reticence, even outright prudery, was much stronger than that of either the working classes or the very wealthy." After all, she said, this was the same group that began referring to parts of a chicken as either light or dark meat rather than saying the words *breasts* or *legs*.

Sleeping on twin beds was one way to paper over the fact that husbands and wives eventually gave in to their basic biological urges. "There was a sense—and I actually remember this from my taking an oral history of my own grandmother—that there was something mildly disreputable about essentially advertising, even to your kids, that you might be having sex together," Coontz said. That prudery and squeamishness lasted well into the 1940s and 1950s. Despite the fact that Lucille Ball and Desi Arnaz were actually married at the same time that they portrayed a fictional husband and wife on television, viewers of *I Love Lucy* saw them nearly every week sitting and talking in their separate twin beds. The only program at the time to show a married couple sharing a double bed was *The Flintstones*. And it featured a yapping pet dinosaur.

Movies weren't much different. In 1934, every major film studio voluntarily agreed to a list of rules that became known as the Hays Code in honor of Will H. Hays, a Presbyterian elder and former postmaster general who took on the role of president of the Motion Picture Producers and Distributors of America. Hays wanted films to be proper moral influences. And under Hollywood's self-censorship, directors had to comply with his code in order to have their movies distributed to theaters across America. When a scene called for a couple to occupy the same bed at once, at least one actor had to keep one foot on the floor at all times to guard against the dire threat of horizontality.

The Hays Code was officially abandoned by the late 1960s, but attitudes toward sex in marriage changed well before that. What seemed modern at the turn of the twentieth century simply felt outdated by the middle of it, in part because baby boomers saw twin beds as something out of their parents' generation. Sex became recognized as not only an obvious part of marriage but also an important part of maintaining a healthy one. Freudian-influenced marriage counselors started worrying about "frigid" wives, and magazines and self-help manuals urged women to become receptive to their husband's sexual needs. Sleeping apart began to be seen as either a sign of a marital problem or something that would eventually lead to one. If a couple wasn't enjoying every moment together—even when those moments conflicted with something as prosaic as sleep—then something was amiss. The pendulum swung back to the shared bed, and for many it took better sleep along with it. "I have taken oral histories of women who mentioned that they had really wanted a separate bed, because their husband snored or thrashed about, but were afraid to ask for fear he would 'take it wrong' or just felt there was something wrong with them for not being able to adjust," Coontz told me.

Attitudes are changing once again, however. It is impossible to know to what extent, but the once-unquestioned idea that relationships are healthy only if a shared bed is involved is weakening just like the dogma of the twin beds before it. Because of busy work schedules, better and more open communication, or the fact that many people wait until they are older to get married and don't want to give up the power of controlling their sleep environment, more couples in happy relationships are choosing to spend their nights in separate beds. As one young physician said, "To be honest, I have never really seen the appeal of spending the whole night sleeping next to somebody. Just because I love someone and want to spend my life with them doesn't mean I want to be in the same bed at the same time. I just don't see the connection." Architects and construction companies surveyed by the National Association of Home Builders predict that by 2016 more than half of all new custom-built homes in the United States will have separate master bedrooms. And yet lingering cultural assumptions make some couples feel like they have to hide it. "The builder knows, the architect knows, the cabinet maker knows, but it's not something they like to advertise because right away people will think something is wrong," one interior designer said about his work designing separate bedrooms for married couples.

Intriguingly, the move back toward separate beds comes at a time when researchers are finding new links between a woman's sleep quality and marital happiness. Wendy Troxel is a professor of psychiatry at the University of Pittsburgh. Early in her career, she noticed that subjects who said they were in high-quality marriages tended to be healthier overall. She began wondering what it was, exactly, about marriages on the less happy end of the spectrum that manifested itself in higher rates of cardiovascular disease and other negative outcomes. Studies had offered theories on stress, smoking, family income, and physical activity. But to Troxel, it seemed like the field was overlooking one of the most obvious aspects of daily life between two people in a relationship. "Sleep was largely neglected despite the fact that we know it's a critically important health behavior," she told me. Even though more than 60 percent of couples sleep with their partner, most studies of marital happiness never considered that it could be a factor.

Troxel recruited couples to wear wristwatch sleep monitors

while they shared their bed each night and to rate each of their interactions with their partner for ten days. When describing each time they had a conversation with their spouse, subjects were given the choice between four positive ratings, such as feeling supported, and four negative ones, such as feeling ignored. Each person in the relationship submitted his or her responses separately, so that a spouse wouldn't feel pressured to modify a rating to appease the other.

The results were clear: the most severe negative ratings came after nights when the woman had slept poorly. Not only that, but the quality of wives' sleep was a more important predictor of happy interactions than a hard day at work or any other form of stress. "Some of that can be because women drive the emotional climate of a relationship more strongly than men in general," Troxel said. "If they have a poor night of sleep they may be more expressive and tend to be more communicative in relationships. A husband is much more likely to pick on his wife's cues that she's had a bad night of sleep than his own."

Men tend to sleep better next to their partners than when they go to bed alone, but that may be because they get to enjoy the emotional benefits of proximity without having to listen to their partner snoring. In one of nature's dark jokes, women not only are far less likely to snore than men but also tend to be lighter sleepers. The result is a nightly farce that is one reason why wives also suffer from insomnia more often than their husbands.

The fact that the importance of sleep is becoming more recognized as a health concern may have the side effect of shaping healthier—and happier—marriages. "One of the values of sleep is that it is a very effective gateway treatment," Troxel told me. "I'm a clinical psychiatrist with a specialty in relationships. In many cases I see patients who would never show up in a general psychotherapy clinic. The idea of sitting on a couch in some therapist's office would go against their entire worldview. But they are concerned about their sleep enough that they're willing to see whomever. And once you get started on sleep, you can address some other issues that otherwise would have been swept under the rug." Returning soldiers, for instance, may be willing to talk about signs of post-traumatic stress disorder if they see it as a way to improve their rest. Because sleep doesn't carry the same stigma that still unfairly lingers for mental health issues such as depression and anxiety, dealing with sleep issues somehow seems both less scary and more practical to some patients. Couples are often willing to change their routine and try separate beds at night if both partners are aware that they are splitting up for better sleep alone, and not because of some unspoken change of heart.

Given that sleep studies consistently find that subjects sleep better when given their own bed at night, why do so many couples decide to deprive themselves of a lifetime of better sleep and remain on a shared mattress? For an answer to that question I tracked down Paul Rosenblatt, a professor in the Department of Family Social Science at the University of Minnesota and one of the few sociologists who has studied couples' sleeping patterns in the United States. He became interested in the topic after what he calls a traumatic experience. A number of years ago he was working on a research project that documented the lives of rural farmers. One family invited him to stay for the weekend, and suggested that he bring his twelve-year-old son along. Rosenblatt readily agreed, thinking it would be a nice father-son bonding experience. But when they arrived at the family's house, he learned that his hosts had only one double bed for the two of them to share. It was the first time that his son had ever spent a night on the same mattress as someone else. "It was hell," Rosenblatt told me. "He had no concept of where his body was in relationship to me. No concept of sleeping on the long side of the bed. By the middle of the night I was clinging to the edge of the bed as if my life depended on it."

Curious after that ordeal, he began looking for academic research on what to him seemed an obvious topic, bed sharing. But out of the more than thirty thousand studies he found that looked at human sleep, couples, or marriage, only nine breached the topic of sharing a mattress. The research overlooked what Rosenblatt considered an important building block in navigating and surviving a relationship. "You learn things by sharing a bed," he told me. "The shock of ending virginity and having sex for the first time is a big deal. But the first time that you share a bed is also a very big deal. Couples can have a very romantic or sexual interest in each other, but if neither has shared a bed before, they are going to have to learn something about getting along—how they spread out, what they do about toenails that are sharp, or if the other person steals a blanket."

He set out to discover why couples opted to share a bed and how the experience affected their relationships. He rounded up couples who lived in Minneapolis and its suburbs, taking care to include subjects spread across the spectrum of love. Some were older and married, some were young and living together, and others were same-sex couples in long-term relationships. Rosenblatt spent several hours interviewing each pair about why they were willing to spend the energy learning to happily coexist on a mattress when it would have been much easier to continue sleeping in separate beds.

The answers were consistent. Couple after couple told Rosenblatt that sleeping in the same bed was often one of their only chances to spend time alone together. If life consisted of playing the roles of parent, employee, or friend, then the shared mattress functioned as a backstage, away from everyday responsibilities and judgments. Bedding down on the same mattress next to a loved one was what made it easier to face tomorrow and the day after that.

That isn't to say that the transition from sleeping in one's own bed to sharing a mattress was an easy one. In one interview, Rosenblatt casually remarked to a subject, a man in his twenties, that it sounded like he somehow learned to swing his elbows less in bed over time. "Not 'somehow,' " the man responded. "There's no 'somehow.' It's her telling me, 'That hurt!' Or, 'Don't do that!' Or 'Watch where you're swinging that elbow!' " Many couples told Rosenblatt that they initially began to share a mattress because that's what they thought everyone did. But, over time, they felt less constrained by their expectations and allowed themselves the freedom to adapt. In another interview, for instance, a couple revealed that one of the most liberating moments of their relationship was when they realized that they didn't have to spoon every night. They could now wake up without sore shoulders, secure in the knowledge that moving to separate sides of the mattress to fall asleep had no greater significance than physical comfort.

Other subjects highlighted the fact that, despite the drawbacks, having another person in bed simply made them feel safer. This was especially true for women. Some female subjects admitted to going to their sister's homes to share a bed rather than face the prospect of sleeping in a room alone when their partner was out of town. Security was also a big concern for older couples. One man told Rosenblatt that he once went into diabetic shock in the middle of the night. His wife woke up, recognized the signs, and called an ambulance. "That is one guy who is never going to want to lie down by himself again, no matter how hot his wife wants the bedroom or whether she likes putting a nightlight on," Rosenblatt said. For these couples, the give-and-take of bedding down on the same mattress was outweighed by a sense of emotional support that could be given only by proximity.

One question still gnawed at me, however. Stanley, the British sleep scientist, argued that there is only one good reason to share a mattress. I asked Rosenblatt about the contention that sleeping in the same bed as one's partner is good for sex and little else. He laughed. If any man actually followed that, Rosenblatt said, he would realize that men who sleep by themselves actually have less sex than those who share a bed with their partners. The change in a couple's sex life after one moves to the room down the hall was so pronounced that men in his study couldn't stop talking about it.

"Some of the men were really grieving the loss of sexual access when they stopped sharing a bed," he told me. "None of the women said that," he added.

The mystery of the shared mattress was solved.

4

And Baby Makes Three

Abigail's bedroom perfectly appeals to the tastes of a two-year-old girl, which, as luck would have it, is exactly what she is. Disney princesses smile down at her from the lilac walls. A small white bookcase sits in the corner, topped with a lamp in the shape of a tulip. If you ask Abigail if she has a favorite pair of shoes—and she's hoping that you will—she will open up her closet, move aside a basket of toys and dolls, and emerge with every single pair that she owns.

In the middle of the room sits a white bed frame, holding a small mattress covered by a comforter with a pattern of daisies on it. This is perhaps the only object in the tidy, small bedroom that she has no opinion of. And why would she? Abigail has never slept in a bed by herself, much less this one, which her parents picked out for her several months ago. For her, the routine of going to sleep means putting on her pajamas, brushing her teeth, and listening to one of her parents sing a lullaby as they put her down in the middle of the king-sized bed in their room. They join her anywhere from twenty minutes to two hours later. It is the same basic script that the family has followed every night since she was born.

Abigail's parents, two white-collar professionals in a major city, didn't mean for their daughter to sleep this way. Before she was born, they bought a cherry-wood crib and spent an afternoon assembling it in what would become her room. Next came the purchase of a white bassinet, which they put in their room and intended for Abigail to sleep in during her first few months, a way station that would make nighttime feedings easier and calm their nerves when she was out of sight. The day soon came when they brought their newborn home. They put her down in the bassinet that first night as planned, but something strange happened as they lay in their bed staring at the ceiling and trying to sleep. Abigail, in the corner of their modest bedroom, felt much too far away. Her father strained to hear her every breath. Her mother wondered if the bassinet was sturdy enough. When Abigail woke up and began crying, both of her parents jumped out of bed together. While her mother held her, Abigail's father dragged the bassinet next to his wife's side of the bed, closing the three-yard gulf that seemed much larger when the lights went down. Finally, with Abigail practically on their mattress, they drifted off to sleep.

At first, they attributed their feelings to first-night jitters. But the next night, they again felt like something was amiss while their daughter slept soundly in the corner. Abigail's father moved her bassinet flush with the side of the bed for the second time. After the third straight night of this, Abigail's parents bought a tiny crib that had barriers on only three sides. The fourth was open except for a small padded rail. Abigail's new space was permanently fixed next to her parents' mattress, an annex that allowed the three of them to effectively sleep in one family bed. She remained in the side-sleeper until she outgrew it. At that point, she began sleeping on the mattress itself, wedged between her mother and father.

Her parents knew they were breaking rules laid down by their doctor, who strongly discouraged the notion of co-sleeping. Not only that, but they were inviting criticism from their own parents as well. Both sets of Abigail's grandparents had been vocal with their thoughts that anyone who slept in the same bed with an infant was negligent. But Abigail's parents had come to enjoy what they saw as an intense bonding time with their child. They went ahead and bought a child-sized bed, knowing that the day will soon come when it will be put to use. In the meantime, they let their parents assume that Abigail spends each night in her own room.

Abigail is one of countless children whose sleeping patterns were far from the mainstream just a generation ago. About one in fifteen parents admitted to sharing a bed with their child in a study published in 1993. By 2007, the number had grown to about one in three. The actual number of *co-sleepers*, a fuzzy term that for some means sleeping in the same bed with their child and for others means sleeping in the same room, might be much higher. Like Abigail's mother and father, parents who sleep in the same bed with their kids can be reticent to admit it, worried that they are going to be scorned by their family, questioned by their friends, and criticized by their doctors.

The growing popularity of co-sleeping troubles many public health officials because the body of an adult can pose a danger to a baby sleeping on the same mattress, especially when that adult has had too much alcohol. Public health officials point to studies such as one conducted in Santa Clara, an upscale California county that makes up most of Silicon Valley, which found that twenty-seven infants over a five-year span died as a result of being placed in the same bed as a sleeping adult. More than half of those accidents were caused when the adult rolled over onto the child. The others were caused by suffocation. The American Academy of Pediatrics warned against co-sleeping in the early 1990s, citing the risk of an infant becoming entrapped in bedding or an adult's clothing.

What accounts for such a dramatic shift in where children sleep despite all of the official warnings against it? It may be nothing more than a newfound willingness on the part of parents to try anything to tame the often-unhappy relationship between children and sleep. Getting children to sleep is the first problem that parents are expected to solve, and yet it is also one of the most difficult. One study found that parents seek advice from their doctors regarding this issue more than any other health concern or behavior.

Part of the reason for the confusion comes down to biology. Infants initially make no distinction between day and night. The inner clock that tells them when it is time to be awake and alert and when it is time to sleep emerges gradually, which means that babies who want to eat or play at two in the morning have no idea why this might be out of the ordinary, and explains why babies who get tired at seven one evening won't necessarily do the same thing the next. During the first weeks of life, infants will typically sleep about sixteen or seventeen hours each day, although the longest stretch will last only about four or five hours at a time. Unlike children who are a few months older, newborns pay little attention to the surrounding environment. Hunger, loud noises, and lights usually won't be enough to keep them from falling asleep when their brain needs it. It is an urge that comes frequently and intensely: infants spend nearly half of the time that they are asleep in deep REM sleep, a level of brain activity that is as busy as when they are awake.

This type of sleep pattern—bouncing between sleep and wakefulness outside of a twenty-four-hour schedule—is known as polyphasic sleep. Sleeping in a more or less single period based on the time of day is called monophasic sleep. Nothing good happens when a polyphasic sleeper comes into a household of monophasic parents. Nighttime feedings may take place at one in the morning one day, and then at three in the morning the next, without any pattern or schedule beyond an infant's capricious cries. Parents have to deal with the consequences of the delayed circadian clock until infants reach about four months old, which is when most children start to sleep in eight- to nine-hour blocks during the night.

Parents aren't quite out of the woods once their children reach toddlerhood, however. While the total amount of time children sleep each night decreases as they get older, their resistance to it builds. Stonewalling becomes a new part of the bedtime routine. Parents soon hear endless requests for another cup of water, for another story, or another song. These nightly struggles take place despite the fact that sleep—whether in long stretches at night or in short doses in the form of naps—is one of the most reliable things that make very young children happy. In one study, researchers identified two groups of three-year-olds. Children in the first group had strict nap schedules that mandated the hours they had to spend in bed, whether they wanted to or not. Children in the second group napped whenever they felt like it, which was rarely. All of the children in both groups slept for a total of about ten and a half hours each night, regardless of whether they napped or not. Yet the children in the dedicated napping group slept more, logging an average of two additional hours over a twenty-four-hour period compared with those with irregular naptimes. The outcome of the extra sleep was better interactions between parents and their children. The children in the napping group were "more fun to be around, more sociable and less demanding," researchers noted. With their longer attention spans and calmer dispositions, they were able to learn and adapt to changing circumstances. Children who didn't sleep as much, meanwhile, were hyperactive and fussy, a result of missing out on the time spent in deep REM sleep that allowed the nappers to better react and respond to the world around them.

No one debates the fact that young children need a lot of sleep. Yet the difficulties in getting them to do it have spawned a mini-economy of parenting books focused solely on sleep, written by a number of competing experts, each of whom claims to know best. Richard Ferber, a pediatrician at Children's Hospital in Boston, wrote one of the landmark books of the field in 1985: *Solve Your Child's Sleep Problems*. Before then, sleep was barely mentioned in the standard child-rearing guides.

Ferber became interested in sleep in the 1970s, shortly after the birth of his own children. As he spent night after night rocking his son to sleep in his arms, only to watch him wake up the minute Ferber placed him in his own bed, Ferber began to wonder why it was difficult for a child to fall back to sleep on his own. He slowly came to the realization that infants simply don't know how to do it by themselves. Gradually, Ferber began weaning his children from what had become the family bed by letting them cry for progressively longer periods of time before he or his wife would check in on them. Ferber hoped that his son would no longer associate falling asleep with being rocked or held and would learn that a parent will not always be available to attend to every one of his cries. Instead, his son would begin to develop the ability to calm himself down. "A baby cannot count sheep," Ferber later told an interviewer. "So we have to find a way to help them. To teach them in a simple, gentle way that they need to sleep. And that they need to do it all by themselves. It really isn't so hard for them, either. Babies love to learn."

The philosophy, which became known as either the sleep-training or the cry-it-out method, became so popular that Ferber's name morphed into a verb. New parents began asking their friends whether they, too, were "Ferberizing" their children, and whether it was working. The drill itself was fairly simple. A parent would place a child down in his or her own bed, and come back to the room at longer and longer intervals to soothe the child. Ferber advised parents to steel themselves against the sound of their children wailing and to stick with the sleep-training plan. With time, a child wouldn't need help. In the first editions of his book, Ferber noted that sharing a bed with a child would likely make the process of developing effortless sleep more difficult. "Although taking your child into bed with you for a night or two may be reasonable if he is ill or very upset about something, for the most part this is not a good idea," he wrote. Parents were also warned that co-sleeping could slow the emergence of a child's sense of independence. "If you find that you actually prefer to sleep with your infant," Ferber wrote, "you should consider your own feelings very carefully."

Part of the appeal of sleep training is that it is designed to allow parents to sleep. In his practice in Boston, Ferber consistently heard from a steady stream of parents about how sharing a bed with their child meant that they never slept for more than an hour or two at a time. They described living in a half-asleep daze, woken up by every cry, and resenting the fact that they felt inadequate at both work and home. This form of chronic sleeplessness has an outsized effect on mothers. One poll of twenty thousand working parents conducted by a team from the University of Michigan found that women are two and a half times more likely to interrupt their sleep to care for a child compared with men. Once a mother is awake, she tends to stay that way for an average of forty-four minutes. When a father wakes up to attend to a crying child, however, he is often able to fall back asleep within a half hour. These moments of male alertness were short and rare. Nearly one out of every three mothers said that they woke up to care for their infants every night. Just one out of every ten men did so. "Obviously, the child-rearing responsibilities maybe slanted at first due to breast-feeding," one of the lead researchers said. And yet "the responsibilities are never renegotiated," she added.

The effects of poor sleep build and quickly manifest in working mothers' lives. Some have difficulties functioning at their jobs, an important concern given that most professionals see their greatest salary increases during their late twenties and early thirties—a time when many working women head home to a young child. The side effects of a crying child in the middle of the night aren't limited to sleepy mothers fighting the urge to nod off at their desks. As one study found, the quality of a child's sleep often predicts maternal mood, stress levels, and fatigue. It's a very simple equation: the more sleep a child gets, the healthier the mother will be.

If Ferber's method was as simple in practice as in theory, then its promise of painless sleep would do a lot to improve the lives of working adults. But it's not simple. The excruciating first nights of the Ferber approach can require listening to a child's searing screams go on for well past what seems safe or healthy. That leads many parents to William Sears, a professor of pediatrics at the University of California, Irvine, School of Medicine, whose approach to sleep is almost the exact opposite of Ferber's. The father of eight children, Sears has become one of the leading voices of what is known as attachment parenting. He believes that through sharing a bed with an infant, parents not only develop a stronger bond with their child but also respond to their needs better. Many parents who subscribe to Sears's approach do so out of the worry that allowing a baby to cry for too long sets in motion a range of long-term health effects. One article in *Mothering* magazine gives a general idea of how far this line of thought goes. "But there is no doubt that repeated lack of responsiveness to a baby's cries—even for only five minutes at a time—is potentially damaging to the baby's mental health," it warned. "Babies who are left to cry it out alone may fail to develop a basic sense of trust or an understanding of themselves as a causal agent, possibly leading to feelings of powerlessness, low self-esteem, and chronic anxiety later in life." James J. McKenna, a professor of anthropology at Notre Dame, has argued that mothers who share a bed with their child are more likely to breast-feed. These babies, when they do inevitably wake up, may also fall asleep faster when their parents are right next to them. With better-quality sleep, the brain would then have more energy to devote to cognitive or physical development.

In many ways, co-sleeping prods parents into reverting to

an approach to sleep that was widely practiced in the United States a few generations ago, and remains common in African-American and Asian-American households today. Until the start of the twentieth century, most American babies were placed in a cradle in the same room as their parents or a live-in nurse. Once old enough, young children graduated to sharing a bed with siblings of the same sex. But, as Peter Stearns noted in a paper published in the *Journal of Social History*, children's sleep habits changed more dramatically between 1900 and 1925 than at perhaps any other time in history. Noisy new inventions like radios and vacuum cleaners entered the home for the first time and gave parents a reason to segregate their children into a quiet place at night while adult life went on. Women's magazines, meanwhile, ran articles written by experts who argued that traditional sleeping habits were dangerous and unhygienic. And if those concerns weren't bad enough, a shared bed began to cause a sort of class anxiety. Middle-class parents, in particular, began to worry that their children's sleeping arrangements said something about the financial condition of the family. Many parents believed that a move out of the city and into the suburbs meant that they had to provide their offspring, even infants, with their own rooms. One sleep expert I spoke with said that some middle-class parents remain adamantly opposed to bed sharing because they see it as a step down the economic ladder, especially if their infant doesn't have his or her own room. "Parents now tell me, 'Oh my God, it's going to be a huge problem that my children are going to have to sleep in the same room,' " she told me. "It's not the question of 'How do I deal with it?' Now it's 'Should I move?' "

In recent years, sleep scientists have begun to join pediatricians and anthropologists in the contested field of children's sleep. What they found may surprise you. Jodi Mindell is the associate director of the Sleep Disorders Center at the Children's Hospital of Philadelphia, the first pediatric hospital in the United States and among the best in the world. There, as part of a team that cares for conditions ranging in complexity from narcolepsy to extreme fussiness, she treats about fifty patients a week. Mindell realized one day that she didn't know the answer to a basic question: how do babies around the world sleep? She could do little more than guess whether parents who put their baby down to sleep in San Francisco did so at the same time or in the same way as their friends in Tokyo.

Along with Avi Sadeh of Tel Aviv University and others, Mindell polled nearly thirty thousand parents of infants and toddlers in Australia, Canada, China, Hong Kong, India, Indonesia, Korea, Japan, Malaysia, New Zealand, the Philippines, Singapore, Taiwan, Thailand, the United Kingdom, the United States, and Vietnam. It was one of the first, and most extensive, surveys of global infant sleep patterns. All of the subjects in the study lived in conditions that roughly corresponded to a middle-class lifestyle in the United States. Each household featured electric lights, televisions, refrigerators, running water, and other comforts. Mindell gave the families a list of basic questions that any parent would be able to answer easily: What times does your child go to sleep? Does your child sleep alone or in a bed with you? And, does your child have a sleep problem?

To say that the answers were unexpected is an understatement. Families on different continents didn't even seem engaged in the same activity. In New Zealand, for instance, the average bedtime for a child under the age of three was 7:30. In Hong Kong, it was 10:30. But bedtimes were not the only difference. Nearly everything that made up the children's sleeping habits depended on their location, a triumph of culture over biology. In Australia, 15 percent of parents said they regularly shared a bed with their child. Almost six thousand miles away in Vietnam, nearly 95 percent of families did so. In Japan, children slept for an average of eleven and a half hours each night. The average infant in New Zealand slept thirteen hours. And, perhaps most surprising, 75 percent of parents in China, a country in which most families are co-sleepers, reported that their children had a sleep problem.

Any hope that a global survey of children's sleep habits could provide an answer to the sleeping-training versus co-sleeping debate vanished. There were simply more variations than researchers thought possible. "I thought that there would maybe be a ten- or fifteen-minute difference in bedtimes and that would be about it," Mindell told me. "Instead we got this eye-opening understanding that sleep is dramatically different in babies throughout the world." She was left with more questions than answers. "We don't know why there are those differences in sleep and what the impacts of them are," Mindell continued. "Maybe someone could argue that Korean babies are getting less sleep and that's because they are going to bed too late. But maybe there's a true biological difference and Korean babies simply need less sleep. That's a very different question and there are a lot of theories out there. It's a whole career to figure it out."

Cultural approaches to sleep work for the most part until toddlers get their first taste of globalization. To illustrate this point, Mindell tells the story of a mother who grew up in England, went to college in the United States, and eventually moved to Hong Kong for work. All of these destinations more or less followed the same Western approach to children's sleep, segregating an infant into his or her own room from an early age. Once in Hong Kong, Mindell's patient hired a nanny to care for her three children while she was at work. The nanny was from a rural area in China and approached each of her charges like she would a child in her own home. That meant that the children didn't go into the expensive crib in the nursery or into their own beds when it was time for sleep, but instead were held in her arms or placed on the mattress next to her. This co-sleeping approach functioned reasonably well during the week. But when Mindell's patient had solo charge of her children over the weekend, the crib regained its starring role. It was a night-mare. The mother couldn't get her children to stop crying no matter what she tried. She asked her nanny to have the children sleep in their cribs or beds, but the nanny refused. After all, she argued, the kids liked it better her way.

At first glance, the point of the story appeared to be that co-sleeping worked better for this family. But Mindell says that wasn't the issue. The children were stuck between East and West, sleeping next to someone one day and sleeping alone the next. It wasn't sleep training versus co-sleeping that was the problem, she says, but consistency. "Children are more likely to be relaxed throughout the bedtime rituals if they have a good idea of what's coming next," Mindell told me. In the case of her patient in Hong Kong, either approach to sleep could have been effective if it was followed regularly.

When it comes to children's sleep, routine is a better predictor of quality than whatever choice the parent makes regarding co-sleeping. Consistently following the same nightly script makes bedtime less of a battlefield. In one three-week study, Mindell investigated the effects of a nightly routine on four hundred mothers and their children, who ranged from newborns to toddlers. During the first week of the study, all of the mothers were told to follow their usual approach to sleep. After that, half of the mothers were given instructions on how to follow a specific plan. Each mother was advised to pick a consistent time that she would place her child in his or her crib or in the family bed each night. Thirty minutes before this bedtime, she was to give her child a bath, followed by a light massage or application of lotion. Then, she was to do a calming activity like cuddling, rocking, or singing a lullaby. Within thirty minutes after the bath, the child was to be in the spot where he or she usually slept, with the lights out. Each mother followed the instructions for two weeks and then reported any changes. By every measure, routines led to calmer nights. Children fell asleep faster, woke up fewer times during the night, and slept longer. When they did get up the next morning, they seemed to be in better moods. Parents improved their sleep quality as well, with the mothers feeling better able to handle their daily challenges.

Mindell's work suggests that the advocates of co-sleeping and those of the cry-it-out method are both a little right and a little wrong. If consistency is the most important predictor of sleep quality, then it doesn't necessarily matter if a child like Abigail sleeps in her family's bed when she is two years old. There are signs that other professionals are softening their dogma when it comes to children's sleep. Ferber, the guru of sleep training, revised his views on co-sleeping in a 2006 update to his best-selling book. He now advises parents that sharing a bed with their children can be a safe and effective option, as long as the parents follow basic guidelines to prevent accidentally harming their infants.

Eventually, almost all children decide to sleep in their own bed when they are given the option. Without prompting, Abigail has begun referring to the bed in her room as her "big-girl bed." Her parents think that it won't be long before she moves out of their bed. But calming their child's ambivalence toward sleep is only part of their job. Soon, Abigail's brain will be developed enough to experience a truly strange aspect of sleep. Abigail, you see, is about to have her first dreams.

5

What Dreams May Come

Alice had lasagna with her dead father last night and is upset that he didn't like the food. She says this while sitting on a metal folding chair in a cramped room in the middle of Manhattan. Outside, the streets are filled with tourists trying to find their way to the Christmas tree in Rockefeller Plaza. Inside, four of us are arranged in a semicircle facing a plastic fern in a bright-blue pot. We have come to this second-floor counseling center on a Sunday afternoon to spend two hours discussing our dreams. Alice is the first up to bat. She lets out a volley of coughs and proceeds to tell us that her father, who died two decades ago, popped up in her dreams several times last week, walking around and criticizing her cooking.

"How did that make you feel?" the woman to the right of me, who is leading the group, asks her.

"Awful. I had planned everything just so," Alice replies.

"What do you think the message of that dream was?" the group leader asks.

"I think that I wanted to tell myself that I wasn't meeting the expectations of my life," Alice responds.

The group nods encouragingly while Alice goes into detail about her dream. I spend the time getting more and more nervous, rehearsing in my head what I am going to say, like an actor reviewing his lines minutes before showtime. I have come armed with two of the few recent dreams that I can remember. The first features a plot that would make for an anticlimactic heist movie. In it, I robbed a bank with three of my friends from high school and then sat eating pretzels in a Florida airport while we waited for our getaway flight. I decided to go with this dream because it was more exciting than the other one, in which I bought a green-and-white cocker spaniel puppy and named him Sprite.

Reciting the dream to a small group of strangers doesn't scare me. It is the fact that these nice people seem convinced that dreams have hidden meanings, and I'm not so sure. The idea that in the middle of the night the brain sends coded messages to itself that reveal deep secrets seems like a plot device out of a bad soap opera. I am of the mind that dreams are more or less random. Though there is no telling whether my view is ultimately the correct one, studies seem to support it. By injecting a solution into a subject's bloodstream that made blood flow visible, for example, researchers found that the brain's long-term and emotional memory centers are most active during REM sleep, the phase of the sleep cycle when most dreaming occurs. That could be one reason why dreams have little narrative cohesion but are laden with moments from the past.

However, the members of the dream group gathered here today would beg to differ. They have come to discuss their dreams because they are convinced there is something inherently important, and even life changing, about their experiences in dreamland. To them, looking at a dream only in terms of the mechanics of the brain misses the point, kind of like basing an evaluation of the *Mona Lisa* on the pH level of the paint used alone. Alice isn't concerned with what part of her brain was responsible for allowing her to interact with her father again. She cares about the emotions she experienced in her dream, feelings that were so strong she remembered them for several days afterward. By definition, that makes them meaningful for her.

The question of whether the contents of dreams tell us anything deep about ourselves presents a dilemma for those who study how the brain works. On the one hand, dreaming is a fascinating biological phenomenon universal to every person and most mammals, as far as we can tell (scientists once tried to ask a gorilla who knew sign language whether she dreamed at night, but the gorilla's attempt to rip the researcher's pants off put a quick end to that). Each night, nearly everyone becomes paralyzed every ninety minutes or so during REM sleep. The brain starts working overtime, and the sexual system perks up.

During this dreaming stage, a man's penis will become erect while a woman will experience increased vaginal blood flow. The brain will then create images and stories that the body responds to as if the events in dreamland were actually happening, as anyone who has woken up sweating and out of breath from a particularly scary dream well knows. These dreams happen regardless of a person's physical state. Those who have lost their sight after they were toddlers continue to dream with images, for instance, while those who were blind from birth dream with sounds. And yet any trance that feels so real during a dream disappears almost immediately upon waking, leading some to believe that they don't dream at all and others, like me, to remember only fleeting pieces that make dreams seem all the more puzzling (a green-and-white puppy?). The fact that all mammals experience dreams in roughly the same way suggests there is something vitally important about this stage of sleep.

Yet here is where the paradox comes in. For professional researchers, announcing that you are investigating dreams goes over about as well as proclaiming that you are intent on finding the lost continent of Atlantis or uncovering a UFO conspiracy hidden by the Federal Reserve. "If you're going to get tenure or make a spectacular career in science, dreams are probably not the thing you want to study," Patrick McNamara told me with knowing understatement. McNamara is the head of the Boston University School of Medicine's Evolutionary Neurobehavior Laboratory, where he studies how the brain reacts in different situations. As part of his work, he has conducted research into dreams, nightmares, and what goes on in the brain during meditations and religious experiences. Even with a professorship and an impressive name for his lab, McNamara detects sideways glances from other neurologists. "Studying dreams is still considered a little New-Agey and not entirely respectable," he said.

No matter its reputation now, the investigation of dreams is one of the foundations of sleep science. Dreams were what drew many early researchers to the field in the first place, driven by the chance to discover the mechanisms and meanings of a nightly experience that has intrigued us since humans scratched out the first written language. Most cultures, and nearly all major religions, have regarded dreams as omens at one time or another. Ancient Greeks thought that dreams were visions given to them by the gods. Early Muslims considered dream interpretation a religious discipline sanctioned by the Koran. And the Bible is a veritable dream fest. In Genesis, God speaks to Jacob in a dream and describes his plans for the Israelites. Later, Jacob's son Joseph interprets Pharaoh's dreams after all of the magicians in Egypt have failed to do so, a feat for which he would later receive a Broadway musical. In the New Testament, a different Joseph gets a visit from an angel in a dream that tells him that his virgin wife is pregnant with God's son and that he shouldn't freak out.

By the start of the modern era, science had become convinced that dreams were essentially nonsense. Yet the suggestion that they revealed something hidden in an individual's mind changed that. In 1900, Sigmund Freud was a forty-three-year-old son of a wool merchant who had a small medical practice in Vienna. That year, he published a book that became the linchpin of dream theory for half a century. In *The Interpretation of Dreams,* he argued that, far from being random events, dreams were full of hidden meanings that were projections of the dreamer's secret hopes and wishes. In effect, Freud identified the subconscious, a realm of thought beyond the mind's control that colors our desires and intentions. Every night when a person went to sleep, Freud said, the mind cloaked these thoughts in symbols that could be uncovered and interpreted with the help of a therapist. Without dreams, our unconscious concerns would be so overwhelming that few of us could function. Dreams were what allowed us to think the unthinkable. These "letters to ourselves," as he called them, were an important safety valve for the mind. Take them away, and psychic pressure would then build and lead to neurosis.

To prove his point, he gave examples of his own dreams. In what would eventually become the most discussed dream in psychology, Freud described seeing one of his female patients among a number of guests in a large hall. He takes her aside and faults her for not accepting his prescribed treatment for her illness. She replies that the pain is spreading to her throat and starting to choke her. He sees that she is puffy and begins to worry, wondering if he missed something in his examination. Freud then takes her to the window and asks her to open her mouth. She is reluctant to do so, and Freud finds himself getting annoyed. Soon, his friends Dr. M and Otto arrive and help him examine the patient. Together, they discover that she has a rash on her left shoulder. Dr. M surmises that the woman's pains are due to an infection, but a bout of dysentery will rid her body of the toxin. Freud and Dr. M come to the conclusion that the cause of the trouble was most likely Otto, who had recently given her an injection of a heavy drug through a syringe that had not been properly cleaned.

On reflection, Freud found this dream much more than a simple, albeit strange, story. "If the method of dream-interpretation . . . is followed, it will be found that dreams do really possess a meaning, and are by no means the expression of disintegrated cerebral activity, as the writers on the subject would have us believe," he wrote. By looking at each aspect of his dream as a stand-in for an emotion or anxiety, Freud found that the dream allayed his concerns that he was responsible for the health of a particularly difficult patient. First, the woman puts up a fight throughout the dream, making it clear that he thinks that any caregiver would have difficulty quickly discovering her problem. This is confirmed when it takes three doctors examining her simultaneously to find the rash on her left shoulder. And with the help of Dr. M, Freud finds that it was Otto who foolishly gave the woman an injection and caused her illness. Taken together, the content of the dream suggests to Freud that he could walk away from his patient, blameless for what happens to her. "The whole plea—for this dream is nothing else—recalls vividly the defense offered by a man who was accused by his neighbor of having returned a kettle in a damaged condition," he writes. "In the first place, he had returned the kettle undamaged; in the second place it already had holes in it when he borrowed it; and in the third place, he had never borrowed it at all. A complicated defense, but so much for the better; if only one of those three lines of defense is recognized as valid, the man must be acquitted."

Wish fulfillment like this could come in many forms in a dream. Freud saw them as a release of anxiety—a condition that he linked with sex, though he described the connection in less-than-direct terms. "Anxiety is a libidinal impulse which has its origin in the unconscious and is inhibited by the preconscious," he wrote. "When, therefore, the sensation of inhibition is linked with anxiety in a dream, it must be a question of an act of volition which was at one time capable of generating libido—that is, it must be a question of a sexual impulse." Perhaps unfairly, Freud's theories soon became reduced to the view that everything in a dream had a sexual meaning that reflected and uncovered long-repressed urges from childhood. One review of Freudian literature found that by the middle of the twentieth century, analysts had identified 102 stand-ins for the penis in dreams and ninety-five symbols for the vagina. Even opposites—flying and falling—were called symbols for sex. Freudians pointed out fifty-five images for the act of sex itself, twenty-five icons of masturbation, thirteen figures of breasts, and twelve symbols for castration.

Freud saw a patient's resistance to this theory of dream interpretation as proof that it was valid. He explained that even he was initially put off by the seemingly absurd notion of his dreams. "When I recollected the dream in the course of the morning, I laughed outright and said, 'The dream is nonsense,'" Freud wrote. "But I could not get it out of my mind, and I was pursued by it all day, until at least, in the evening, I reproached myself with these words: 'If in the course of a dream-interpretation one of your patients could find nothing better to say than "That is nonsense," you would reprove him, and you would suspect that behind the dream there was hidden some disagreeable affair, the exposure of which he wanted to spare himself. Apply the same thing to your own case; your opinion that the dream is nonsense probably signified merely an inner resistance to its interpretation.' "

The fact that Freud didn't interpret the dream about his patient along psychosexual lines spurred a subschool of analysts devoted to unlocking additional meanings from that dream alone. In 1991, for instance, a paper in the *International Journal of Psychoanalysis* postulated that the dream actually reflected the fact that "Freud may have been haunted by the repressed memory of an incident of erotic aggression enacted by himself against his sister Anna when he was 5 years old and she 3 years old."

The Freudian view of dreams held considerable sway among psychologists well into the early 1950s despite complaints that the theories were too focused on sex. In one scientific journal, a critic wrote, "We have seen that a multitude of symbols can stand for the same referent. Why is it necessary to have so many disguises for the genitals, for sexual intercourse and for masturbation?"

Freudian analysis became a popular part of culture by the 1920s, influencing everything from movies to the study of crime. William Dement, a professor at Stanford University who is considered one of the deans of sleep science, was attracted to the field in the 1950s because of the chance to immerse himself in the Freudian study of dreams. "There was a belief that Freudian psychoanalysis could explain every aspect of our problems: fears, anxieties, mental illnesses, and perhaps even physical illness," he wrote.

But it was Dement, in part, who helped science lose an interest in dreams. As a medical student at the University of Chicago in the early 1950s, Dement began some of the first systematic studies of REM sleep. This stage of sleep had been discovered only in 1952, when researchers in a laboratory at the same university believed that a malfunctioning machine created the appearance that a sleeping subject's eyes were moving rapidly during the middle of the night. Unable to detect the cause of the problem, the researchers decided to go into the room and shine a flashlight on the subject's eyes. They found that the eyes were in fact darting back and forth under the eyelids while the body lay still. This realization unearthed the fact that there were different stages of sleep. After finding that subjects woken up from REM sleep were the most likely to remember their dreams, Dement organized studies of this stage of sleep in infants, women, and those with mental illnesses, all in an attempt to see if time spent dreaming shed any light on Freud's theories. "It is hard to convey how exciting it was to be doing this work," Dement wrote in his memoir. "Here I was, a mere medical student, holed up in a nearly deserted building making one surprising discovery after another . . . I imagine that this was how the first man to discover gold in California must have felt in 1848."

Dement's discovery that the brain is as active during REM sleep as it is when a person is awake transformed sleep research. Here was a period of brain behavior that was unlike anything else. Dement proposed that science should recognize that the human brain rotates through three distinct time periods: sleep, awake, and REM sleep. Other researchers initially scoffed at the idea. Dement's paper on the subject was rejected five times before it was published. "People reacted as if I were claiming that we don't need air to breathe," he later wrote. But his theory soon became an accepted fact, and led the view that REM sleep is perhaps the most important of any of the sleep stages.

Other experiments revealed how odd this dreaming stage of sleep truly is. In France, a researcher named Michel Jouvet called it paradoxical sleep because the body was immobilized while the brain looked to be fully awake. He then conducted one of the most famous experiments in sleep science. By making small lesions in a tiny part of a cat's brain stem known as the reticular formation, Jouvet discovered that he could block the mechanisms that normally suppressed movement during REM sleep. The result was that he could watch the animals act out their dreams. While fully asleep, these cats arched their backs, hissed, and pounced on unseen rivals. The behavior "could be so fierce as to make the experimenter recoil," he wrote. Once a cat pounced onto an object with enough force, it would jolt itself awake and look around dreamily, surprised at how it got there.

In a strange way, Jouvet's discovery that it was possible to know exactly what a cat was dreaming about made the content of human dreams a lot less interesting to researchers. Once dreams could be identified and recorded by brain waves, they no longer seemed a mystical, complex reflection of the human subconscious. The dreaming stage of sleep was soon identified in almost all birds and mammals, lessening the importance of human dreams by comparison. As Jouvet later wrote, when describing why neurologists had turned away from dream studies, "What significance is there for a newly hatched chick to realize any desire other than to become a cock or a hen?" Eventually, researchers found that human babies in the womb also cycle through REM sleep—and presumably dreams.

REM sleep largely split the field of sleep from psychology, as neurologists moved to incorporate this stage of sleep into a better understanding of the brain. Freudian interpretations of the meaning of dreams still continued, but mostly on analysts' couches. In research labs, however, the content and possible meanings of dreams were largely put to the side and ignored.

Dream research remained stagnant until a psychology professor at Case Western Reserve University in Cleveland named Calvin Hall decided to catalog what people dream about. Hall spent more than thirty years gathering dream reports from everyone who would share them. By the time he died in 1985, Hall had synopses of more than fifty thousand dreams from people of all age groups and nationalities. From this large database, he created a coding system that essentially treated each dream like it was a short story. He recorded, among other things, the dream's setting, its number of characters and their genders, any dialogue, and whether what happened in the dream was pleasant or frightening. He also noted basics about each dreamer as well, such as age, gender, and where the person lived.

Hall introduced the world of dream interpretation to the world of data. He pored through his dream collection, bringing numbers and statistical rigor into a field that had been split into two extremes. He tested what was the most likely outcome of, say, dreaming about work. Would the dreamer be happy? Angry? And would the story hew close to reality or would the people in the dream act strange and out of character? If there were predictable outcomes, then maybe dreams followed some kind of pattern. Maybe they even mattered.

Hall's conclusion was the opposite Freud's: far from being full of hidden symbols, most dreams were remarkably straightforward and predictable. Dream plots were consistent enough that, just by knowing the cast of characters in a dream, Hall could forecast what would happen with surprising accuracy. A dream featuring a man whom the dreamer doesn't know in real life, for instance, almost always entails a plot in which the stranger is aggressive. Adults tend to dream of other people they know, while kids usually dream of animals. About three out of every four characters in a man's dream will be other men, while women tend to encounter an equal number of males and females. Most dreams take place in the dreamers' homes or offices and, if they have to go somewhere, they drive cars or walk there. And not surprisingly, college students dream about sex more often than middle-aged adults.

Hall's research deflated the idea that dreams are surreal.

The plot may not follow any logical order and characters may have strange requests, but the dream world isn't that far from reality. More important, dreams tend to be unpleasant. Hall found that the average dream is filled with characters who were aggressive, mean, or violent. Dreamland, in short, sounds a lot like the worst days of middle school.

The discovery that dreams are often negative perked the interest of neurologists. Why would we have dreams if most of them tended to be unhappy? Are our brains depressed novelists? The answer, some said, comes from imagining the purpose of dreaming in the context of evolution. In a 2009 paper, a Finnish cognitive scientist named Antti Revonsuo argued that negative, anxiety-filled dreams were simply an ancient defense mechanism, letting us experience bad things in order to train our brains to react in case something similar happened while we are awake. Dreams, in this view, are the brain's dress rehearsals. For evidence, Revonsuo pointed to data collected by Hall in which the dreamer is running away from something or being attacked. "Because adaptations presumably require hundreds of generations to change, currently living individuals still carry those adaptations that were designed to work in the ancestral environment, regardless of whether the adaptations serve their original functions in the radically different modern setting," Revonsuo wrote. In other words, our ancient ancestors likely had negative dreams before a planned hunt or battle. Today, dreaming of getting attacked is how the brain prepares for the anxiety of a big sales presentation at work, and there is nothing we can do about it.

A problem with this theory is that not all unpleasant dreams are of the I-am-being-chased variety. Take, for instance, the dreams of a man named Ed, who kept a journal of his dreams about his wife, Mary, for twenty-two years after her death. Ed and Mary met on a boardwalk in 1947, when he was twenty-five and she was twenty-two. She died of ovarian cancer three decades later. When Ed dreamed about Mary after she was gone, the plot often followed the same theme: Ed and Mary started off happily engaged in an activity before something happened that split them apart. Sometimes the stories were full of cinematic images. In one dream, for instance, Ed sees Mary sitting in a car across the road, but he can't find a way to reach her. At other times, Ed's dreams introduced absurd elements into vignettes of everyday life, such as the time when Ed and Mary bump into Jerry Seinfeld and ask him for directions. Before Ed knows it, Seinfeld has walked away with Mary, and Ed has been left alone. Ed goes behind a building to brood, where the ground beneath him turns into quicksand. Individually, all of the elements of the dream are recognizable from daily life. But add them together, and there isn't a clear and present danger the brain is preparing for.

I am able to tell you the details of Ed's dreams because of G. William Domhoff, a professor at the University of California, Santa Cruz, who collected dream journals alongside Calvin Hall and made their vast collection available to other researchers in the early 1990s. As he read countless accounts of dreams, Domhoff began to see that most people had dreams like Ed, featuring the same situations and characters for years on end.

With enough reports from any one person, Domhoff argued, the stories in the dreams can give an accurate reading of the dreamer's concerns, all without having to resort to Freud's idea of interpreting symbols. Just look at Ed, he said. He had recurring dreams in which he was separated from the love of his life for two decades after her death. It doesn't take an analyst's couch to realize that he missed her.

I spent a sunny afternoon with Domhoff in Santa Cruz, sitting in a Yogurt Delite on the Pacific Coast Highway and talking about dreams. "None of Freud's claims are true by any of our standards today," Domhoff said, dipping his spoon into his yogurt. "If you look at dreams—if you really look at them like we have—then you see that it's all there, out in the open. You don't need any of these symbols." He went on. "Freudians got all caught up in the idea that there were hidden meanings to our dreams. But their interpretations only worked because we share a system of figurative language and metaphor."

As an example, he had me imagine that I had a fairly straightforward dream. "Let's say you have this dream that you are on a bridge to an island and then the bridge starts to shake and then you run back. Now what would you say if I asked, 'Now, what do you suppose that bridge is a symbol of?' "

"I don't know," I said, with my mouth full of yogurt.

"But you do," Domhoff replied. "You have a metaphor system. You'll 'cross that bridge when you come to it.' It's a transition. So I say that you're in the midst of a transition. But we're *all* in the midst of some sort of transition in our lives. Then I can say that the dream means that you are afraid to take the next step. You want to stay with solid land instead of going onto the island. It all makes sense because I've assumed that the dream is metaphoric and I'm giving you a metaphoric interpretation. If I stay very general all of this works. Now that I know a little bit more about you, I can guess and get more specific. I can say that the island means that you're writing a book and going out on your own. I've got a plausible interpretation. But really I'm just a fortune-teller going from a lot of clues."

Looking at a history of a single person's dreams reveals that the brain doesn't construct such clear metaphors, he said. Instead, dreams are filled with images and settings that are familiar. If a woman dreams about walking over a bridge, it is more likely that she literally crosses a bridge during her daily commute, or that she can see one from her window at home, than that her brain has decided to broadcast her emotions using figurative images.

Freud, on the other hand, thought that a stranger dream signified a deeper meaning. In *The Interpretation of Dreams,* he argued that "dreams are often their most profound when they seem the most crazy" because they were more densely packed with symbols to unlock. I asked Domhoff how, even if they don't happen as often, seemingly surreal dreams, such as of flying or getting trapped in a strange room, could possibly serve as a mirror to our daily lives and concerns.

He decided to answer my question with a story. One woman who sent him her dreams gave herself the code name Melora. Hall and Domhoff had asked the subjects to obscure their name before contributing to the dream bank. Melora, for those who don't know, is the name of a character from a well-known episode of *Star Trek: Deep Space Nine,* a television show that ran in the mid-1990s. Domhoff's subject chose the code name because she was a big-time Trekkie who loved reading science fiction. On a less pleasant note, she was also a mother going through a divorce. In most of her dreams, she is concerned about her child or she is spending time with her ex-husband, doing such routine things as hiking together or spending a holiday at his parents' house. But every once in while, she is charging through the cosmos. "Sure, sometimes she has a fantastic adventure, but she lives in that world because of her enormous amount of reading and love of science fiction," Domhoff said. Just like work or family, *Star Trek* was another part of her life that showed up in her dreams. Trying to parse the meaning of why one of Melora's dreams took place aboard a spacecraft and another took place in her office couldn't tell you anything, Domhoff said. But, taken in context of hundreds of other dreams, the small number of intergalactic dreams reflected the fact that science fiction was important to her. The things that you care about are the things that you dream about.

That is not to say that Domhoff thinks there are great meanings or evolutionary advantages lurking in dreams that have yet to be discovered. Dreams are just "an accidental by-product of our ability to think and have an autobiographical memory," he said matter-of-factly. We dream about negative things, in Domhoff's opinion, simply because we spend a lot of time worrying. The easiest way to see this in your own life is to start a new job. For the first week or so, there is a good chance that your new commute, new coworkers, or new responsibilities will take center stage in your dreams. In many of these dreams, you will probably disappoint yourself or others in some way. Students during the first week of school often dream about getting lost on their way to class, for instance, while waiters dream about dropping food or spilling wine down a customer's shirt. "Dreams are worst-case scenarios that reflect what we think about every day," Domhoff said. "We take all these little could-be's and we blow them up." In real life, that is what our minds do with many of our problems, anyway. Dreams could be the manifestations of the brain taking our anxieties and running with them because there is nothing else competing for its attention in the middle of the night.

Ernest Hartmann, a professor at the Tufts University School of Medicine, agrees with Domhoff that the content of dreams matters, but with a slight twist. Hartmann sees dreams as a form of built-in nocturnal therapy. In dreams, he says, the mind takes what is new or bothersome and blends it into what the brain already knows, making the new information seem less novel or threatening. In what I have unscientifically come to think of as the Well-Adjusted Caveman Theory, Hartmann argues that the life of early man was filled with the kind of traumas—watching friends gored by animals with sharp tusks or fall through holes in the ice and drown, just to give you two possibilities—that few people experience today. Those who were able to regain their emotional balance after living through a traumatic event were more likely to survive over the long run than those who dwelled on the negative.

As evidence of this theory, Hartmann points to the fact that the mind has a tendency to replay scary or harrowing experiences in dreams almost exactly as they happened in real life for several nights after the event. It is not exactly impartial to judge the validity of a theory by your own experience, but this point resonates with me. The summer after college, I was on a rural one-lane dirt road in the woods of Northern California, riding in the passenger seat of a friend's Mustang, when a white Bronco sped around a blind turn right in front of us. The Bronco swerved to the left to avoid a head-on collision, but it wasn't quick enough. It rumbled over the hood of our car like a monster truck, missing my seat by inches. We were lucky. I wasn't hurt, and my friend walked away with only a broken arm. The car, however, was practically totaled. For the next week or so, I woke up sweating from nightmares of still being stuck in that passenger seat, watching as the Bronco's tires grew menacingly larger amid the sound of crumpling metal.

I don't remember exactly how long it took to have a dream that wasn't about the accident, but the nightmares eventually went away and never amounted to anything more than a couple nights of poor sleep. For some people, however, the brain gets stuck replaying traumas, like a band that knows only one song. When the brain fails to set aside the event in its long-term memory—a move that researchers see as a sign that the emotional system has come to accept what happened and can now put it into perspective—a person may experience recurring nightmares, which is one of the hallmarks of post-traumatic stress disorder. Some grow to fear sleep.

Since at least the Vietnam War, when more than one in five combat veterans returned home with chronic nightmares, drugs have been the main line of defense against a brain stuck in a cycle of bad dreams. But there may be a better way. Doctors currently think it is possible to train the brain to dream about other subjects and characters, in a sense rewriting the stories that we tell ourselves each night. One promising technique is called imagery rehearsal therapy, a two-step process in which patients first describe the traumatic event or person that continues to reappear in their nightmares. Then, they choose a situation or image to replace it. Before going to sleep, patients spend at least ten minutes thinking about the dream they want to have, positioning themselves as the director of the show rather than an audience member. Over time, this appears to work. In one study of combat veterans, imagery rehearsal therapy was as effective at reducing nightmares as medication.

Domhoff calls this form of therapy a clear advance in the science of dreaming because it does not imply that the mind is masking its concerns in symbols or storylines. Nightmares themselves don't seem to serve any function other than to frighten us, he argues, and we may have more of them than we realize—we simply forget about them shortly after we wake up. What Domhoff calls the clearest example yet that Freud's dream interpretation theory is misguided is the lack of evidence that people who remember their bad dreams are more in touch with their emotions than those who don't. Dream theory in the twenty-first century is pointed toward uncovering anxieties, not symbolism. Psychologists now look to what dreams can do for us—our understanding of ourselves, or how a soldier copes after returning from fighting in Afghanistan—instead of whether dreams have an intrinsic meaning or represent a suppressed urge.

That concept closely resembles what I experienced at the dream group that cold day in Manhattan. Alice wasn't concerned about her possible repressed feelings when she was searching for meaning in the dream in which her dead father criticized her. In fact, she sounded a lot more like Ed, the man who spent two decades dreaming about missing his wife. Just like Ed couldn't let his love for Mary go, Alice couldn't come to peace with her father's memory.

A few months later, I returned to the dream clinic, curious to see whether Alice would still be there and what she was dreaming about. I walked up the stairs to the counseling center and made my way down the hallway. I was the first one to arrive. Over the next twenty minutes, seven people filled in the seats around me. But still no Alice.

I had a report ready this time. In the dream, I received a call at work from my college's registrar's office. The office had recently audited the files of past students, and found that I was six units short of the requirements to complete my degree. I was told that I would need to take two classes that summer on campus or else I would have to make up the full four years again. The fact that I was living in New York at the time, and not Southern California, made this impossible. In the last part of the dream, I was frantically calling the dean's office to explain my situation. Domhoff, if I had asked him about it, would most likely tell me that this is nothing more than an example of my mind going through a worst-case scenario and that I should ignore it.

Alice walked in with a few minutes to spare. The session started and we went around the circle, introducing ourselves. Alice was the second person to share a dream. I leaned in, wondering what was going to come next. Was she going to tell us about her father again?

"So I had a very strange dream that I wanted to share with the group," she began. In her dream, she told us, she was babysitting her grandson when he suddenly disappeared from her couch. She searched the house and couldn't find him. Her cell phone began to ring, and just when she was about to answer it, she was woken up by the sound of her phone ringing in real life.

"How did the dream make you feel?" the session leader asked.

"Like a bad grandmother. I swear, all my dreams end up this way."

6

Sleep on It

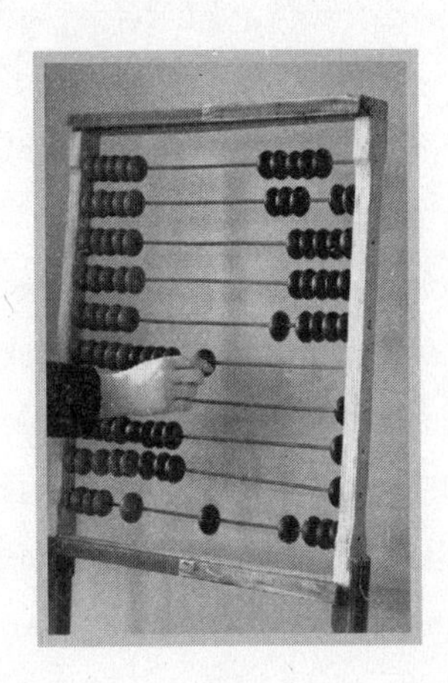

In the early 1960s, Jack Nicklaus was in the midst of a remarkable run. At age ten, he had picked up a set of golf clubs for the first time and promptly won the juvenile tournament at the Scioto Country Club, a prestigious club in Columbus, Ohio. Competing as an amateur at age twenty, he had placed second at the U.S. Open, finishing only two strokes behind Arnold Palmer, then the world's best golfer. Nicklaus had won the U.S. Open the year that he began golfing full-time as a pro, and won the Masters and the P.G.A. Championship the next year. In two years, his ability to sink a ball into a hole had earned him what in 2011 would now be worth $1.2 million.

Nicklaus entered the 1964 U.S. Open as the favorite. He had just finished second in the Masters behind Palmer and was looking to even the score in their building rivalry. As he walked onto the course on a hot and humid day at the Congressional Country Club in Bethesda, Maryland, Nicklaus was counting on the choreographed power of his swing—which artfully combined brute force with graceful accuracy to smash a ball far onto a fairway—to put him ahead of the pack on one of the longest courses in the tournament's history.

On the first hole, he plunked the ball directly into a sand trap. It was the first sign that something was off. Fourteen times over the course of the day, he used his driver—normally his best club—to aim for the fairway from the tee. His ball landed on the fairway of only six holes. The rest of the time was spent hacking his way out of the rough, out of sand traps, and out from behind trees. After the first round, Nicklaus was tied with thirteen other golfers at a score of 72, four strokes behind Palmer. The next day was even worse, with Nicklaus ending with a 73, and the day after that was worse still, with Nicklaus scoring a 77. By the end of the final round, Nicklaus had finished the tournament he was supposed to win tied for twenty-third place. He walked home with a measly $475.

Swinging a golf club is one of the hardest skills to learn in any sport. There are just so many ways to mess it up. Turn your hands in too much, and you will watch the ball sail off anywhere but the place you were aiming for. Bring your arms around at the wrong time, and the powerful stroke you were envisioning will do nothing more than plop the ball five feet in front of you.

Connect the face of the club either too high or too low on the ball, and you could be left with a stinging vibration in your arms and a ball that is buried in the grass. Professional golfers are so much better at the game than everyone else because they have trained their bodies to rotate at the right speed while keeping their arms at the right angle, again and again, without variation.

Somewhere on the course in Bethesda, Nicklaus lost that fine sense of timing that was his livelihood, and he didn't have long to get it back. In three weeks, he was scheduled to play in the British Open at St. Andrews, one of the most challenging courses in the sport, and the tournament where he had lost by a single stroke the year before. Bookies in London, expecting him to regain his form, listed him as the favorite to win. But what they didn't know was that as he went through the dynamics of his swing, Nicklaus couldn't pinpoint what was wrong. Every golfer has a bad day, or even a bad week. But Nicklaus's troubles were showing no signs up letting up. The poor performance at Bethesda lingered past the event, and if it continued, it could put his career in jeopardy. It was as if the extra amount of talent that had separated him from the rest of the pack had vanished without warning. After years of routinely shooting under 70 strokes per round, Nicklaus found himself in the unfamiliar position of readjusting his expectations downward.

A few nights before he planned to leave for Scotland, Nicklaus went to bed still puzzling over what had happened to his swing. That night, he dreamed that he was once again pounding the ball onto the fairway, regaining his form as the Golden Bear pictured in so many sports sections. When he woke up, he suddenly realized that he had held the club slightly differently in his dream, an adjustment that allowed him to keep his right arm steady throughout the swing. It was a tweak that would be barely perceptible to anyone else, but Nicklaus instantly recognized that this was the cause of his recent troubles. He got out of bed and went directly to the course. There, he gripped the club like he had in his dream. He shot a 68, and the next day he shot a 65. His old stroke was back. "Believe me, it's a lot more fun this way," he told a newspaper reporter. "All I had to do was change my grip a little." He went on to finish second at the British Open after scoring one of the lowest rounds in the long history of the tournament.

In his sleep, Nicklaus had figured out a problem that was costing him ten strokes a round, which, in the high-stakes world of professional golf, was the difference between walking home with a $100,000 check and barely making enough money to cover the airfare. His dreaming brain was able to do something that it couldn't when Nicklaus was awake and studying his poor performances. Clearly, something happened that night that led Nicklaus to wake up with a solution to his swing problem. But what?

For as long as we have been dreaming, the stories our minds create while we are asleep have been credited as a source of insight. As Nicklaus's dream about his golf swing shows, the breakthroughs aren't always emotional. When a person lies down to sleep at night, the brain undergoes a process that is crucial to learning, memory, and creativity in ways that scientists are only now beginning to understand. What dreaming does for our brains is most evident in stories like Nicklaus's, in which the mind solves a problem or develops a new thought without any conscious effort.

Scientists and others who investigate how the mind works have long attributed insights like these to flittering strokes of genius, a mysterious dance of cells and neurons which adds up to a thought that changes the game. Viewing creativity and problem solving as onetime events echoes the thinking of the Ancient Greeks, who believed that ideas came from the Muses and that you needed to work to win their favor. Even hardened scientists with no inclination to believe in mythology have been amazed at how the mind, while in a dream, sometimes suddenly reaches the perfect solution to a problem. In 1865, August Kekulé, a German chemist, was working on a model of the structure of benzene, an important industrial solvent whose chemical makeup had been confounding engineers and scientists at the time. Kekulé woke up from a dream with a vision of a snake eating its own tail. As he lay in bed, he realized that benzene's chemical bonds would fit into the same hexagonal shape. The discovery was so important to German industry that Kekulé was awarded a title of nobility. Albert Szent-Györgyi, a Hungarian scientist who won a Noble Prize in 1937 for isolating vitamin C, credited his dreams for regularly revealing solutions to stumbling blocks. "My work is not finished when I leave my workbench in the afternoon," he wrote. "I go on thinking about my problems all the time, and my brain must continue to think about them when I sleep because I wake up, sometimes in the middle of the night, with answers to questions that have been puzzling me."

Naturally, dreaming and its role in creative thought are more celebrated the farther you move away from the hard sciences. In perhaps the most famous instance of dreams leading to art, Samuel Taylor Coleridge awoke from an opium-induced dream in 1816 with three hundred lines of poetry in his head. He was in the middle of writing them down when he was interrupted by a visitor, who stayed for nearly an hour. When Coleridge returned to the poem, he could remember only fragments of what had appeared so vividly to him in his dream, which accounts for why the last stanzas of his masterpiece "Kubla Khan" seem disjointed. About 150 years later, Paul McCartney woke up in his girlfriend's bedroom thinking of a melody. He went straight to a nearby piano and began playing the tune for the future hit "Yesterday." "It was just all there," McCartney later told a biographer. "A complete thing. I couldn't believe it." In the summer of 2003, Stephenie Meyer was a stay-at-home mother living in the Arizona suburbs. On the day she was supposed to take her children to their first swimming lessons, she woke up from a dream in which a girl was talking in a meadow with a beautiful vampire, who was trying to restrain himself from killing her and drinking her blood. She immediately wrote down the conversation from the dream as accurately as she could remember it. That dream became the basis for the *Twilight* series of books and movies, which have since earned Meyer more than $100 million.

On the surface, it seems like these ideas came out of nowhere. But a tiny bit of excavation shows that each dream had clear connections to what was happening in that person's daily life.

Complex, creative thoughts that appear fully formed were little more than solutions to life's problems. Kekulé had been searching for the structure of benzene for months. McCartney was part of one of the most productive songwriting duos in history and in the midst of creating a historic succession of hit records, but he was facing the fact that the next Beatles album needed another song to be complete. Meyer had been starting and stopping ideas for novels for years, trying to find characters that were real enough to hook readers.

Dreaming looked to be the time when the mind continued to work in its laboratory, testing approaches and solutions to situations that were a part of its waking life. But how? And could the process of creativity really be tested and observed in a research lab?

In the 1960s, leading psychologists turned their attention to how we develop innovative solutions to problems. To begin with, these researchers had to define what they meant by creativity. The working definition they came up with was "the forming of associative elements in new combinations which either meet specified requirements or are in some way useful," a definition wide enough to include both the chemical structure of benzene and the tales of lovesick vampires. The next step after defining creativity was to see whether there was any replicable method for how the mind comes up with new ideas. Psychologists crafted a four-step model to chart how we typically react when faced with a new problem that has no easy or obvious solution. In the first step, we engage in an intense but unsuccessful session in which our minds grapple with the basic elements of the problem or issue. Then we tend to put it aside and focus on other things that require immediate attention. That leads to a dormant period in which the problem doesn't take up any conscious thought or attention. Finally, the solution comes to us in a sudden flash of insight at a time when either we are not thinking about it or we are dreaming.

The most important part of the puzzle is what happens in the brain between when we put a problem aside and when a solution flashes in front of us. Is it simply the passage of time that allows the brain to come up with the new idea, or is there something more at work? In the early 1980s, Francis Crick and Graeme Mitchison theorized that dreaming was a crucial element of learning and creativity, two closely related skills that can lead to advantages for survival, ranging from finding scarce food to creating a new product for a business. That sleep—especially REM sleep—could be a time when the mind solves a problem makes intuitive sense. REM sleep is when our most vivid dreams occur and a period in which the mind is as active as it is while awake. If we spend little time in REM sleep one night, our brain will compensate by prolonging that stage of sleep the next night. It doesn't take a huge leap to assume that the brain considers this time important.

According to Crick and Mitchison's theory, the brain picks up countless bits of information throughout the day, from the structure of the face of a waiter at lunch to the color and pattern of a coworker's tie. When we learn something new—whether it is declarative, such as the facts of what happened at work last Wednesday, or procedural, like how to drive a car—

the information flows through a part of the brain called the hippocampus. Storing all of this information into long-term memory not only is impractical but also could slow our brains down from finding something important when we need it. The brain picks and chooses what it keeps and what it tosses, so that information that isn't essential is forgotten to make way for what is coming the next day. The process of cleaning up and organizing the mind's filing cabinet could take place during REM sleep, which would account for the randomness of dreams. Touches of creative genius are simply exaggerated versions of what happens when our brains remove the clutter every night. With only important information left, the mind may then be free to make associations that it couldn't see before.

While this theory was intriguing, Crick and Mitchison didn't do much to prove it. In the first part of the 2000s, a team of researchers at the University of Lübeck in northern Germany decided to put it to a test in a laboratory setting. The question they hoped to solve was whether sleep was the catalyst for a new idea, or whether the time the brain spent working through a problem accounted for insights. They assembled a group of volunteers and asked them to solve a number puzzle. Researchers explained to the subjects that, to reach the six-digit answer to the problem in front of them, they should follow two rules that required no math skills beyond basic subtraction. The first step was to look at the relationships between six pairs of numbers in a string of digits. If a subject saw something like two 4's in a row, he or she was told to respond with the repeated number. But if the two numbers were different, then the correct response would be the difference between them.

What the researchers didn't tell the subjects in the study was that there was actually a much easier way to get to the answer. In each instance, the second three digits in the answer were the mirror images of the first. That meant that if the first part of the answer was 4-9-1, the second part would be 1-9-4. It was a subtle pattern that no subject recognized during the training session, even after completing a block of thirty trial runs.

After everyone knew how to solve the puzzle the long way, the researchers broke them up into groups based on how many hours they would get to sleep. One group was allowed to sleep normally for eight hours. Another was kept up all night. The third group, subjects who were trained to solve the puzzle during a morning session, was asked to come back eight hours later without taking a nap in between. Through this setup, researchers ensured that each group stepped away from the problem for the same number of hours. If the groups more or less improved equally, it would suggest that solutions to problems come after the brain has a long enough time to reflect. But if the improvement rates between the groups were different, it would suggest that something happened during sleep and dreaming that made a difference in their ability to interact with new challenges.

When the results came in, it was clear that sleep was key. Subjects who did not get to sleep before their second shot at the puzzle showed little improvement. Those who slept eight hours, meanwhile, solved the task 17 percent faster. But that wasn't all. The subjects who figured out the hidden, easy solution to the puzzle completed each set approximately 70 percent faster than their peers because they had to solve only the first three digits in the six-digit answer. Only one out of every four subjects in the groups that did not get to sleep caught on to the pattern by the end of the study. But almost everyone who slept eventually discovered the quick solution. Sometime in the night, their minds were able to construct a novel approach to a problem they had faced while awake. Subjects who didn't get to sleep continued to conceive of each puzzle literally, following the by-the-book instructions handed to them by the research team. Sleeping, meanwhile, allowed the brains to develop a cognitive flexibility that led them to consider the situation in a new way.

It was as if sleep stretched the muscles of the brain, and it responded by bending its conception of facts and reality in a way that let it arrive at a new vision. While this study confirmed that sleep did in fact enhance problem solving, the question remained of whether dreaming played a role in the process. Were dreams just a part of sleep that occurred at the same time that the brain was consolidating its memories and honing its new skills, or did dreams help the brain reach its goal?

Back across the Atlantic one researcher at Harvard University turned to video games in his investigation into how the brain tags new information that later reappears in dreams. Robert Stickgold, a professor of psychiatry who was then in his early sixties, became interested in dream studies because of an experience he had had while hiking with his family in Vermont. One night, as he started to drift off to sleep, he felt like he was still on the mountain. Even though he was comfortably in bed, he had the very real sensation that he was grabbing rocks and pulling himself up. When he woke up two hours later, the feeling was gone.

A few days later, he mentioned to his colleagues that he had the strange sense that his mind was replaying its day just as he fell asleep. He then learned he wasn't alone. His friends told him that they had had the same experience after completing intense, focused activities like whitewater rafting, or—this being a group of Harvard professors—studying organic chemistry all day. Stickgold wanted to conduct a study to see whether this was a common occurrence, but he was stuck trying to design an experiment that didn't require his bussing subjects to Vermont and leading them up a mountain.

That's when a colleague suggested Tetris. One of the most popular video games in history, Tetris requires players to sort falling pieces of assorted shapes into straight lines while listening to the soundtrack of a Russian folk song. As anyone who has played the game knows, there is something about it that sticks with you when you are sleeping. Stickgold assembled a group of college students, taking care to include those who had never played the game before and those who had spent more than fifty hours on the game. As part of the study, Stickgold let the subjects fall asleep normally in rooms in his sleep lab. He woke them up not long after and asked what they were dreaming about. Approximately three out of every five replied that they saw falling Tetris pieces. The challenges that the brain had grappled with during the daytime replayed in the mind as the subject went to sleep, just like with Stickgold's sensation of climbing over rocks after his day in Vermont.

More reports of Tetris dreams came on the second night of the study. It seemed that once the mind realized that being asked to sort falling shapes wasn't a fluke, it decided to devote extra time to figuring out a strategy. All of the subjects who were new to Tetris reported seeing game pieces in their dreams, while only half of the experts did. Intriguingly, Stickgold included in the study several subjects who regularly suffered from amnesia. Among this group, too, he received reports of dreams of falling shapes, even though the subjects could not consciously remember playing the game. Each person's brain used sleep as a time to rehash what it experienced while awake. When subjects played the game a second time, their Tetris dreams appeared to help them improve more than simply time alone.

Other studies showed the same thing. Researchers in Brazil, using the violent first-person shooter video game Doom instead of Tetris, recruited volunteers to play the game in which they blasted zombies and monsters with shotguns, knives, and chainsaws for at least an hour before they fell asleep. When they were woken up from REM sleep and asked what they were dreaming about, monsters and chainsaws topped the list. Just like with Tetris, subjects who spent more time dreaming about the game demonstrated a greater improvement in their skills the next time they played than those whose brains hadn't relived its experiences during sleep.

Across town from Harvard, at the Massachusetts Institute of Technology, a neuroscientist named Matthew Wilson found that new information a rat learned during the day was incorporated into its dreams as well. He implanted tiny electrodes into the brains of his test subjects. He then recorded each rat's brain waves as the rat searched through a maze. Wilson focused on a cluster of neurons in the hippocampus that were responsible for storing memory, including memories indicating that a particular place contained food or was difficult to maneuver around—a job that is very similar to what the hippocampus does in our own brains. While the rats slept, Wilson noticed that the pattern of their brain waves almost perfectly matched what he saw while the rats were awake and moving through the maze. The data were so similar that Wilson was able to tell exactly what part of the maze the rat was dreaming about. The animals were replaying what they went through during the day and committing it to memory.

Just as Crick and Mitchison proposed, sleep appeared to be the time Wilson's rats focused on new and important information. Stickgold decided to take this line of study further, using the next best thing to electrodes implanted in human heads: more video games. Thanks to the Tetris experiments, Stickgold convinced Harvard to buy an arcade game for him called Alpine Racer 2 and install it in his sleep lab. The game was part of a new line of machines that required players to move their whole bodies, rather than just their thumbs. To play the arcade game, someone steps onto two platforms, each of which represents a ski, and grasps two movable blue handles, which stand in for poles. Players must move their legs and arms simultaneously to dodge trees and slalom through gates in a rough approximation of tackling a black-diamond run in Colorado. The immersive experience was a clear parallel to hiking in Vermont or any other full-body activity that melds decision making with physical movement, a taxing cognitive process in which time and patience lead to skill.

Stickgold designed a study that would test whether humans continue to dream about new information throughout the night. His goal was to determine how novel data interacts with what the mind already knows. Like in the Tetris experiment, Stickgold recruited volunteers, who then played the game for forty-five minutes and slept in his lab that night. But this time, he decided to wait until some subjects had completed one or two sleep cycles, the roughly ninety-minute loops that the brain goes through every night, before he would wake them up and ask what was going on in their dreams. As in the Tetris experiments, almost half of the subjects who were woken up early in the night had dreams that seemed out of the video game, dreams of skiing or hiking in the mountains. But as the night progressed, the dream reports became less straightforward. Subjects began to say they were dreaming of things like moving quickly through a forest as if on a conveyer belt.

The literal replay of new information had started to evolve into analysis. Once an initial phase of dreaming passed, the brain began finding connections and associations with the data embedded on its memory cards. This stage of dreaming, which fused elements of skiing with what the subjects already knew, occurred later in the night, a time when the adult brain spends longer amounts of time in REM sleep. As the subjects slept, their brains conducted free-association sessions, desperately searching for connections. That may explain not only why the dreams we remember upon waking up after a long sleep seem so strange, but also how we craft new ideas from our memory. The open interplay of emotions, facts, and fresh information allows our brains to see things in a new way. A golfer waking up with a better way to grip a club suddenly seemed less like a genius, and instead the natural outcome of what sleep does for the brain.

Stickgold's contention that the brain consolidates information during sleep in order to make new connections was supported by research conducted by one of his former students, a red-headed Englishman named Matthew P. Walker, who is a professor at the University of California, Berkeley. Working off of Stickgold's research, Walker decided to look at how sleep affected what is known in neuroscience as brain plasticity, which is essentially the way the brain remolds and updates itself when it learns a new skill or stores a new memory. At the time, Walker was fresh from a postdoctoral study at Harvard. He had been part of a team that found that subjects who were tested on their ability to type a string of numbers completed the task 20 percent faster when they were given a chance to sleep before approaching it a second time.

In his work at Berkeley, Walker asked right-handed subjects to type a five-number sequence using their left hands. It was an unfamiliar task that lowered the chances that a subject could skew the data because of his or her natural ability. By analyzing the time that it took them to hit the keys, Walker found that almost all of the subjects subconsciously broke the string of digits into smaller, easy-to-manage chunks, much like you might remember your Social Security number by slicing it into a group of three, two, and four digits. You can hear this process at work when you say your number aloud and find yourself breaking into a singsong rhythm. Walker then had his subjects come back after a night of normal sleep. Just like in the studies of Wilson and Stickgold, time spent sleeping improved performance. After eight hours of sleep, nearly every person in the study typed the numbers in one smooth motion.

Not all sleep gives the brain the same benefits, however. Timing matters. The smoothing effect Walker identified depends on the quality of sleep a person gets immediately after learning something new. The most important period of learning occurs in the first six hours of the night. In one study, researchers trained subjects to perform a motor-skill test. One group was awakened after less than six hours of sleep and trained to perform a second, unrelated task. The other group was allowed to sleep normally. Subjects who did not have their sleep interrupted were able to complete the motor-skill test by an average of 21 percent faster the next day. Those who were awakened, however, improved by an average of only 9 percent. Their brains, it appeared, were interrupted at a crucial time.

In another study, researchers trained subjects to complete a typing exercise. Some subjects were deprived of sleep for one night while the rest were allowed to sleep normally. The next night, however, every person in the study was allowed to sleep as long as he or she wished. By the end of the several-day study, all of the subjects had slept roughly the same total number of hours. Yet even when cumulative sleep hours were more or less equal, a subject's performance clearly reflected how many hours he or she slept the night after learning the new skill. Those who didn't get to sleep that first night consistently lagged behind those who did. The brain's initial shot at consolidating new information into memory mattered more than the simple passage of time, or extra sleep on a second night. The adage that practice makes perfect looked to be only half right. Success depended on practice, plus a night of sleep. "Sleep is enhancing that memory so that when you come back the next day you're even better than where you were the day before," Walker says.

But what if you aren't always able to get a full night's sleep? After all, many of us sleep fewer hours than we would like to after picking up important information—say, what a client expects to spend on our company's products next year, or how to use an expensive new computer program—not because we want to but because we are forced to. Does knowing that we have deprived ourselves of a vital time for learning and innovation make it all the worse?

Not necessarily. If you can't get in a full night's sleep, you can still improve the ability of your brain to synthesize new information by taking a nap. In a study funded by NASA, David Dinges, a professor at the University of Pennsylvania, and a team of researchers found that letting astronauts sleep for as little as fifteen minutes markedly improved their cognitive performance, even when the nap didn't lead to an increase in alertness or the ability to pay more attention to a boring task. Researchers at the City University of New York, meanwhile, found that naps helped the brain better assess and make connections between objects. Test subjects in their study were shown pairs of objects and told that they would be tested on their ability to remember them later. One group was given a ninety-minute break to take a nap, while the other group spent that time awake watching a movie. Subjects came back to the testing room expecting to complete the simple memory puzzle. Researchers instead asked them to describe the relationships between the objects that made up each pair, rather than recall the pairs. Again, the amount of time that each subject slept determined how he or she performed on the task.

Subjects who were able to reach deeper stages of sleep demonstrated a better command of flexible thinking, a vital cognitive skill that allows us to apply old facts and information to new situations. Sleep can also help the brain to recognize patterns. In a study conducted by Simon Durrant, a professor at the University of Manchester in England, subjects listened to a short burst of melodic tones. Later, they were asked to recognize when this sequence played during a much longer stretch of music. As in the study by the New York researchers, subjects were divided between those who were able to take a nap and those who weren't. The amount of time spent in deep, slow-wave sleep during the nap predicted a subject's later performance. Those who reached deep sleep in the nap had a greater rate of improvement at the task of recognizing the abstract patterns in music than did those who simply stayed awake.

Subjects who have been allowed to take naps have finished mazes faster, have become less emotional when confronted with disturbing images, and have remembered a longer list of words than their peers who hadn't been allowed to doze off. Scientists currently think that the process of clearing the mind of unnecessary information and honing skills is at work in all stages of sleep. The benefits improve exponentially the longer someone rests.

Naps are even being used to provide a competitive advantage in the workplace. Companies such as Google, Nike, Procter & Gamble, and Cisco Systems have installed designated napping areas in their offices. The idea is that naps may allow engineers and designers to arrive at creative solutions more quickly than they would by staying awake all day. Consultants from companies with names like Alertness Solutions charge thousands of dollars to educate corporate managers and their employees on the importance of sleep and managing fatigue levels while on the job. Only this time, it is not about worker safety. It's about speeding up the process of manufacturing ideas.

There is one line of work where spending more time sleeping is now seen as a critical part of not only generating creative ideas but also saving lives. It only took a few accidents on the way to get there.

7

The Weapon "Z"

As the sun fell over the desert, the rumble of tanks slowed and then stopped. The men of the Second Armored Cavalry Regiment found themselves surrounded by a sea of sand. It was February 25, 1991, and about fifty hours into the Gulf War, or what would later be called the Hundred Hour War. In that short amount of time, an international coalition led by the United States had pushed the Iraqi Army out of Kuwait and had it on the run. Throughout the Arabian Desert, Iraqi forces found themselves stuck in slow, outdated tanks that Saddam Hussein had purchased from the Soviet Union. Not long after the sale, the Kremlin realized that the tanks had serious problems with firing accuracy and rushed to upgrade the Soviet fleet, but Saddam apparently hadn't kept the receipt. Iraqis soon discovered another curious fault of Soviet engineering: when hit, the turrets of their tanks broke cleanly apart from the vehicle and shot into the air, a phenomenon that coalition forces dubbed the "pop-top." All across the desert, the tops of Iraqi tanks lay smoldering in the sand.

After racing into southern Iraq, the Second Armored Cavalry was told to wait. Another coalition unit had pushed ahead of schedule, and plowing along with them meant there could be a chance of friendly fire. The tanks formed a literal line in the sand and waited in the dark desert night, battered by high winds and heavy rain. It was their first real break after days of combat. Over each of the last five nights, the men had slept less than three hours. Nevertheless, they stood fixed at their stations with night-vision goggles strapped to their heads, staring into the blankness surrounding them. At one in the morning, the first few blips of activity appeared in their sights. The Iraqi forces that unknowingly corresponded to those movements had no idea they were about to run into a row of armed Americans who had the ability to see in the dark. As soon as the U.S. forces confirmed that the vehicles ahead of them were the enemy, they reverted to the unambiguous motto of all tank crews in battle: if it's ahead of us, it dies.

The night air echoed with the sound of gunfire. During the fight, three tanks on the far right side of the U.S. line were forced out of the pack. They took evasive maneuvers, turning left and right to regain an upper hand. Suddenly, the number of enemy forces seemed to triple. Stuck out of position, the American tanks fired round after round from their twenty-five-millimeter cannons. They destroyed the vehicles in front of them and survived without taking a hit. By the time the shooting stopped, the American forces had destroyed every Iraqi tank, but they had also lost two of their own.

Later, in the briefing room, military planners wondered how Iraqi forces were able to zero in on tanks that were faster, stronger, and equipped with better technology. Throughout the whole war, enemy forces would destroy only about two-dozen American vehicles. Yet here, in a small battle conducted entirely in the middle of a rainy night, there were two down. Was it bad luck? Or were some Iraqi forces equipped with a weapon that the U.S. military didn't anticipate? A team went to examine the wreckage of the two doomed Bradleys and interviewed their crews, all of whom survived thanks to their fire-retardant suits, which could withstand temperatures up to two thousand degrees. Sifting through the burning metal, investigators soon held the answer: casings from antitank rounds that could have been fired only from a U.S. vehicle. The two tanks had been lost to friendly fire.

It was a scenario that kept playing out across the desert. In the Gulf War, one of every four American combat deaths was a result of fire from U.S. forces. Not long after the war ended, a team of U.S. Army psychologists began to investigate why soldiers kept attacking the wrong people. Like polio or smallpox, friendly fire should have been eradicated by technology and training. In the months leading up to the Gulf War, tank commanders had spent hundreds of hours in simulated battles.

Crews inside each tank had laser-guided sensors that could identify a vehicle based on the heat it gave off, while soldiers on the ground carried six-pound packs containing receivers that grabbed information from orbiting satellites so they could pinpoint nearby coalition forces on a map. The fog of war, if not entirely dissipated, should have shown signs of burning off. But the rate of friendly-fire accidents wasn't falling like it was supposed to.

Something was happening in the midst of battle that soldiers weren't prepared for and that was leading to casualties. Researchers interviewed soldiers who shot at their own forces and those who were the targets. They pored over training manuals. They built intricate timelines, precise to the second, that detailed when the mistaken shots were fired and what was going through each soldier's mind at the time. They compared real-world conditions with results gleamed from research studies, considering everything from reaction times to morale.

After all of their digging, one truth stared at them, a conclusion that was as obvious as it was radical: soldiers simply weren't getting enough sleep. The skills and training built up over hundreds of hours of preparation were lost on the battlefield amid the sleep deprivation of combat. The effects of sleep deprivation were strong enough that they threatened to derail the greatest military organization in the world. The needs of the human body, and the vital role that sleep plays in how the brain makes rational decisions, were trumping the top-secret technology and hardware that should have given U.S. forces total dominance over their enemies.

Men and women in the military fly around the world and do everything but sleep. Their lives are set out to the minute. In combat zones, most soldiers have no say in when they wake up, when they eat breakfast, or when they lie down at night. In peacetime, soldiers will be lucky if they get six hours of rest a night, or about three-quarters of what most adult bodies need to maintain an alert brain. Adolescent bodies—including the thousands of young recruits who are still not old enough to drink legally—often need nine hours of sleep to be fully restored.

Without deep sleep, the brain morphs from being our greatest evolutionary asset to our greatest weakness. During a study of crew members on a U.S. Coast Guard ship traveling from Virginia to Nova Scotia, for instance, researchers found that twelve of fourteen sailors fell asleep at their posts at least once. It would be impossible to put a dollar cost on all of the bad decisions resulting from the hours of lost sleep, but here is one small but telling number: in 1996, a time of relative peace, crew fatigue was blamed for thirty-two accidents that destroyed American military aircraft, including three F-14 jetfighters that cost $38 million each.

Twelve years after the Gulf War, American tanks once again rumbled through the desert of southern Iraq, this time with the purpose of going all the way to Baghdad. War planners had taken into account how much fuel, food, and ammunition each unit would need for Operation Iraqi Freedom to succeed. Sleep, however, wasn't considered a necessity. The result was an army full of soldiers who had slept only two hours a night during the countdown to war. The dash north from the border of Kuwait stalled several times when drivers behind the wheels of tanks and Humvees fell asleep and veered off the side of the road. The lack of sleep "is our biggest enemy," one marine colonel said during a break in combat. "It makes easy tasks difficult."

Officers in combat nevertheless pushed their soldiers to go without sleep, willfully trading the tactical advantages of speed and mobility for the drawbacks built up from fatigue. Sleep deprivation increased up the chain of command. Many officers stayed awake for more than forty-eight hours straight during the first stage of combat, and when they did sleep, they would lie down for only twenty-minute naps at a time. One commander said that he was able to function after sleeping a total of two hours spread over several days because he was "actively fearful of screwing up."

But fear of failure can take you only so far. Because of rampant sleep deprivation, most men and women in uniform rely on stimulants to stay awake. Chief among them is caffeine. Soldiers guzzle the stuff, starting in boot camp. As they move up the ranks, most graduate from high-caffeine, high-sugar drinks such as Red Bull, Jolt, and neon-green Mountain Dew and turn to super-caffeinated coffee. One popular brand, Ranger Coffee, mixes Arabica coffee beans with additional liquid caffeine. The result is so potent that each bag comes with a label warning that it is not for the faint of heart.

Caffeine is popular as a stimulant because it readily crosses the barrier between the blood and the brain. Once in the brain, it blocks the absorption of adenosine, a nucleotide that slows down nerve connections and makes you feel drowsy. The result is like being able to drive a car backward to roll back the odometer. In research studies, caffeine helped sleep-deprived subjects better discriminate between colors, sort words according to their meaning faster, and see in the dark more clearly. The effect is so powerful that some soldiers resort to eating frozen coffee grounds to stay awake in the field. In the late 1990s, military researchers received a quarter of a million dollars to develop caffeinated gum as an alternative. Gum is ideal if you really need caffeine in a hurry: it allows the stimulant to be absorbed through the tissues in the mouth and reach the brain about five times faster than a pill or coffee. By the 2001 invasion of Afghanistan, pieces of gum that packed 100 milligrams of caffeine—a little bit more than a shot of espresso at Starbucks—were a standard part of soldiers rations. Packets are currently available to civilians on Amazon.com, each one stamped with the hard-sell slogan "Stay Awake, Stay Alive."

When caffeine doesn't do the trick, pharmaceuticals take over. Soldiers in combat have turned to amphetamines since World War II. The military restricts most soldiers from taking the pills—which are essentially speed—without a doctor's prescription, but they are common among certain occupations. Pilots, for instance, routinely take orange "go pills" before night missions, and sometimes take another dose while in the cockpit. The surge in energy comes with a cost. In addition to making it harder to achieve deep sleep once the effect of the stimulant wears off, taking amphetamines can lead to increased aggression and paranoia. Go pills were cited as a contributing factor in a 2001 accident where two American pilots dropped a bomb on an elite Canadian army unit during a live-fire exercise in a remote area of Afghanistan, killing four of them.

Modafinil, marketed to consumers as Provigil in the United States and as Alertec in Canada, is the newest drug found in a soldier's medicine cabinet. Though scientists aren't sure exactly how the drug works on the brain, it appears to increase serotonin levels in the brain stem. Some who have taken the pill reported staying up for thirty straight hours without a noticeable drop-off in ability. But evidence suggests that one danger of the drug is that a user doesn't realize the effects of sleep deprivation. In research studies, sleep-deprived soldiers who downed modafinil were overconfident in their abilities for several hours after taking a dose. The surge in confidence led them to becoming blasé about taking risks they might have otherwise avoided.

They were more fun to be around, however. In one of the high points of military research, army psychologists decided to test whether the use of stimulants influenced the ability to detect and appreciate humor. Getting a joke is tougher than it looks. In the milliseconds between seeing or hearing something and recognizing that it is funny, the brain goes through complex forms of higher thought, such as recognizing patterns, understanding abstract concepts, and appreciating gaps in logic. Test subjects were kept awake for forty-six hours and then shown a series of cartoons and newspaper headlines as part of the Humor Appreciation Test developed by the University of Pennsylvania. Subjects who were given modafinil scored significantly higher than those who drank coffee, suggesting that the drug improved their cognitive performance.

No drug or procedure has been found to replicate and replace the benefits of sleep. It is unlikely that there ever will be. The Defense Advanced Research Projects Agency—the Pentagon division responsible for the invention of the Internet and the stealth bomber—concluded as much in 2007 after many tries. Its goal was to develop a way for a soldier to go without sleep for one hundred hours and still perform common tasks. The military spent millions of dollars testing theories, such as whether it would be possible to put half of the human brain asleep at a time, essentially allowing a person to sleep like a dolphin. None of the tests worked. The only way to recover from lost sleep was to get more of it later.

The invasion of Iraq prompted the U.S. military to rethink the way it approached sleep. The public reason given for the change was that the service needed to keep enrollment numbers up to meet the demands of fighting two wars. Drill sergeants were instructed to spend less time yelling at recruits in boot camp and more time talking with them about their personal goals. At the mess hall, soldiers suddenly had the option of dessert after most meals. Sleeping periods were extended by more than an hour, with lights out at 9:00 p.m. and wake-up at 5:30 a.m. "It has been great for morale," one drill sergeant said at the time. "A soldier's happiness is directly proportional to the amount of sleep he gets."

But the extra sleep wasn't only about comfort. While technology has given the United States a distinct advantage in war, the human body has remained essentially the same. Our brains haven't progressed at the same rate as our technology, which means that the computer on a nuclear submarine is taking orders from a solider whose mind is designed to hunt and gather. When an early human was sleep deprived, the greatest risk was that his prey would get away and he would have to return home empty-handed. Now, with weapons at a soldier's fingertips that could literally destroy the world, the risk is much greater. But to lower the chance that even highly trained soldiers will make mistakes, the military had to understand exactly how sleep interacts with our ability to come to a reasonable conclusion.

If someone asked you why you decided to do something—put on a green shirt today instead of a blue one, become an accountant instead of a sailor, marry your college sweetheart instead of the dancer you met in a café in Barcelona—you could probably give an answer that combined some elements of emotion and reason. But for a long time, these dual tracks of decision making puzzled scientists trying to sort out the process of how we make choices amid the limitless options we are given in life. Plato, among the first to examine the way the brain arrives at a decision, likened the rational part of the brain to the driver of a chariot, and the rush of emotions that we experience to his horses. When impulsive feelings pulled against the brain, it was the job of the driver to rein them in and provide direction. "If the better elements of the mind which lead to order and philosophy prevail," he wrote, "then we can lead a life here in happiness and harmony, masters of ourselves." Letting our emotions take over, meanwhile, would result in ending up "like a fool into the world below."

The concept that the mind was divided into an emotional and a rational half took hold in Western culture. Philosophers from Descartes onward noted the struggle between reason and feeling and envisioned a world in which logic kept us out of pain. While this philosophy was useful in telling us how we ought to think, it came with two problems. The first was that the chariot drivers of the mind were not doing the best job, because humans were not evolving into a coolly logical species. We continued to make choices that were not purely rational, leaving us much closer to the hot-tempered Captain Kirk than the emotionally detached Spock. Clearly, emotions play some kind of role in decision making or else we wouldn't keep turning to them. Second, all of the theoretical ideas of how the mind balanced conflicting impulses didn't address the way the brain literally worked.

Reason—which, along with the inclination to wear pants, is what separates us from our pets—had to come from somewhere. Scientists began the search for the spot in the brain responsible for rational thought. One decided to tackle the problem by cutting out parts of a monkey's brain to see what happened. Monkeys that had their temporal lobe removed showed no fear or anger and tried to eat anything put in their mouths. From this, researchers realized that some small parts of the brain were responsible for higher thought and the regulation of emotion, and that damaging or removing them would drastically alter the way the brain perceived reality. For a monkey without a temporal lobe, everything looks like a banana.

The schematic design of the brain eventually became clear. A bean-shaped structure near the exact center of the brain called the thalamus lets us realize when we are sleepy, while its neighbor, the hypothalamus, monitors feelings of hunger and thirst. Groups of neurons about the size of almonds that are found near the ears, called the amygdala, are partly responsible for the formation of memories, especially those generated by an emotional experience. The nearby pituitary gland and adrenal cortex, meanwhile, send urgent messages in the form of hormones throughout the body when something frightens us.

What regulates all of the messages from the different parts of the brain is a mass right behind the forehead called the prefrontal cortex. Like a conductor in an orchestra, this part of the brain strives to hit the right balance between responses from the emotional parts of the brain and those from the areas responsible for higher thought. The outcome is a decision. The prefrontal cortex is working every waking second, directing attention in the supermarket, sustaining interest when balancing a checkbook, and suppressing any outward signs of frustration or anger. It notices patterns, and when something novel pops up, it goes to work assessing how new information gels with what the brain already knows. It is responsible for a wide range of decisions, both conscious and unconscious, from the recognition that the person walking toward the car is your brother to whether investing in a condo in Phoenix is a good idea.

Making decisions is a very taxing job, with no downtime. Unlike other parts of the brain, the prefrontal cortex gets no benefit from the time that the body spends in a relaxed environment. Even when you are swaying in a hammock sipping a cool beverage on a sunny afternoon, this part of the brain is constantly on alert, making sure you don't topple over or spill your drink. While science still doesn't know exactly how this happens, the time we spend in deep sleep is when the prefrontal cortex recovers and reboots for the next day's work.

In 1999, Yvonne Harrison and James Horne, two professors at Loughborough University, an institution on the far outskirts of London, decided to test how sleep deprivation affects the ways the brain reacts to changing conditions. They developed a computer game that reflected the ebb and flow of the business world and found a number of MBA students to serve as test subjects. Just as they would in their future careers, each student was asked to promote sales of a hypothetical product until it achieved market dominance and profitability. Unbeknownst to them, the dynamics of the imagined marketplace changed halfway through the game once more competitors selling similar products appeared. Suddenly, strategies that used to work made sales plummet. Only students who recognized that they needed to change their strategy would be able to survive.

Harrison and Horne split the students into two groups. Those in the first group were able to sleep as much as they wished, while those in the second had their sleep restricted. Students who slept well watched their sales suffer when new competitors first entered their imagined marketplace, but most were able to recover quickly and adapt. Their counterparts didn't fare nearly as well. After thirty-six hours, the sleep-deprived students were unable to cope with the unseen changes in the game. They continued to rely on what had worked before, not recognizing that these moves now cut into their bottom line. Soon, each was bankrupt. Without sleep, their brains lost the ability to consider any alternatives and became rigid in their logic. It was as if the mind's conductor had forgotten the symphony and focused solely on the oboe, without noticing that something was amiss. Later brain scans showed that after as little as twenty-four hours without sleep, the neurons firing in the prefrontal cortex started to slow down, making it harder for us to complete a thought or see a problem in a new way.

That sleep deprivation makes the prefrontal cortex less adept at realizing the meaning of new information coming into the brain presents a problem when one is formulating a business strategy. Sometimes, however, new information takes a more threatening form, like the appearance of eight Japanese warships. That was the case just after midnight on August 9, 1942, when Allied troops were in the midst of their first major offensive battle in the Pacific theater during World War II. An invasion force of over eighty Allied ships had descended on Guadalcanal, a strategic outcrop of land about a thousand miles northeast of the tip of Australia. Marines had secured the island the day before, and the sailors manning the battleships ringing the island were on the lookout for a Japanese counterattack. Men on duty had been on antiaircraft watch—an alertness level just notches below full combat—for three straight days. They were exhausted, hungry, and sleep deprived.

The attack came just after one in the morning. Japanese battleships steamed toward the American fleet sitting near Savo Island, a speck of land twelve miles off the shore of Guadalcanal. The attack fleet ran head-on into the USS *Patterson,* which radioed a report of enemy ships approaching to the USS *Vincennes* and USS *Quincy,* two heavy cruisers not far away. The captain of the *Patterson* maneuvered his ship into position and gave the order to launch torpedoes, but the men below didn't respond in time. The Japanese fleet sped by and then split into two. Twelve minutes later, one portion of the fleet caught the *Vincennes* and the *Quincy* off guard, despite the warning they had received from the *Patterson.* Meanwhile, the other group of Japanese ships pounced on the USS *Astoria,* another heavy cruiser. The ship's captain woke up to the sound of the general alarm ringing and the rumble of cannon fire. Dazed and still thinking that any attack would come from the air, he ordered his men to cease firing out of fear that they were shooting at an Allied ship. It was a decision that cost lives.

The Japanese forces continued their assault and soon sunk the ship. Throughout the nighttime battle, four Allied ships were lost, killing more than a thousand men and leaving another seven hundred wounded. It was one of the most embarrassing defeats ever suffered by the U.S. Navy, and it generated a minor scandal. Military planners and politicians wanted to know how it was possible for the Japanese to sink so many Allied ships without much of a fight. After learning that he would be singled out in an unflattering navy report, the captain of one battered ship committed suicide.

Fifty years later, a navy psychologist named Nita Lewis Shattuck (formerly Miller) came across a description of the battle. A career military researcher, Shattuck had long studied human performance and the ways that the design of a ship workstation could affect sailors' abilities to complete their tasks under stress. Ships had been updated since World War II, of course, but she hoped to find clues as to whether the layout of each ship contributed to the defeat. After reading only a few pages, she recognized that sleep deprivation played the biggest role that night in the Pacific. With so little sleep after days of combat and the need to maintain a constant state of readiness, the men aboard those ships simply weren't able to react to an attack that came in an unexpected form. Their brains could not shift their cognitive framework from scouring the skies to patrolling the waters, missing the boats in plain view because they had their minds locked on the idea that enemy planes were the greater danger. To an overworked brain expecting an attack from the air, any ship on the water—even those firing at you—could not be the enemy. Like the MBA students who went bankrupt, the soldiers showed the signs of a prefrontal cortex that was exhausted and unable to respond to a situation that had changed.

The role of sleep in allowing the brain to adapt to new circumstances was noted in research studies well before it was fully understood. As early as 1959, the U.S. military noticed that the lack of sleep could severely undermine the discipline of troops going through routine chores. After keeping a group of soldiers awake for a couple of days, one military researcher noted, with the air of a disappointed parent, that "several participants passed through periods of giddiness and silly laughter, like addled drunks, when their behavior became uninhibited."

In the early 1980s, the military began studying how specific tasks were affected by a lack of sleep. The results were troubling if you wanted to win a war. After a day and a half without sleep, air force bombers changed their vocal patterns, no longer enunciating or speaking loudly enough for their crewmates to always understand them. All of the nonverbal clues indicating that something was important—like raising one's voice to suggest danger—vanished from their speech patterns. In another simulation, researchers separated pairs of soldiers into different rooms with a radio linkup between them. One soldier was given a featureless map with only a route and a destination drawn in, while his partner was given a map with all of the basic landmarks present but nothing in the way of directions. In order to get anywhere, the two would need to work together. The pairs made up of soldiers who had slept well completed the task with little problem. Those deprived of sleep for forty-eight hours were another story. The lack of sleep diminished their ability to communicate, effectively wiping out any sort of spontaneous dialogue—the running chatter that helps keep everyone focused on the same goal. It was rare for a pair of sleep-deprived soldiers to piece together a complete map, an essential communications task that soldiers who had slept well accomplished easily. The same breakdown occurred in another study that followed a crew of army soldiers operating on drastically reduced sleep during a simulated battle. Fighting exhaustion, they forgot to do critical tasks such as updating maps with new information and completing boring but important chores. The problems only worsened as the simulation progressed.

In every situation, the prefrontal cortex—the only part of the brain that has the power to think about how it is thinking—had lost the vital aspect of self-assessment, unable to tell if an action was helping to solve a problem or simply making it worse. Without sleep, the brain's finely tuned mechanics had dissolved from an orchestra led by a conductor into a room full of musicians playing to their own beats. In a report warning that sleep deprivation could lead to mission failure, Shattuck argued that "the ability of sailors to think and reason while in a fatigued state has significant implications for combat effectiveness . . . [I]n such a scenario, sailors may give everything they have to give for the mission, but due to human physiology and as a result of fatigue brought on by sleep disruption, their best may not be good enough. The end result of a flotilla of sailors holding key operational positions, all operating in severe sleep debt, could be disastrous."

Military disasters can take many forms. As the nature of war changes—the epic battles between sophisticated national military forces in World War II have now been replaced by door-to-door searches for rogue enemy combatants in the mountains of Afghanistan—the decisions that each soldier makes resonate far beyond the battlefield.

In the first few months of 2007, the United States had a precarious hold on Baghdad. One March evening, a band of American soldiers from the 172nd Infantry came under attack while driving through the streets of the capital. They returned fire and eventually chased four men into a warehouse. There, they found a weapons stash that included several machine guns, grenades, and a sniper rifle. The four men were handcuffed, and the convoy turned and headed toward a makeshift jail under American control. A few minutes later, their superior's voice crackled on the radio. There wasn't enough evidence to keep their new prisoners locked up, he told them, and then ordered the men released.

The order was never followed. A few days before, a roadside bomb had killed two of the unit's men. The soldiers resented the fact that they had risked their lives apprehending four men they believed to be insurgents, only to be told to let them go. Three officers—including the unit's medic—decided to take their prisoners to a canal that snaked its way through an industrial area in a remote part of town. There, they ordered the men, blindfolded and with their hands tied behind their backs, to line up against the back of their vehicle. The Americans pulled out their nine-millimeter pistols and shot each in the back of the head. They dumped the bodies into the canal and drove away.

At a court-martial in Germany two years later, the Americans who were on patrol that night stood trial for murder. Each had admitted to shooting the captives. They pled not guilty, however. Their lawyers said that they were so sleep deprived they could not make a rational decision. In calling the men's actions a regrettable but common part of war, one of the attorneys said that "good soldiers who freaked out in the field of battle largely as a result of sleep deprivation and a lack of battlefield backup are spending a lot of time in jail." A military psychologist, too, testified that sleep deprivation could have played a role in the shootings. It was not enough. All four soldiers were found guilty and sentenced to twenty-year terms at military prisons in the United States.

In one way, the shootings can be seen as a failure of the prefrontal cortex. As with the sleep-deprived men who seemed to turn into drunks in the army study, the soldiers' emotions and impulses were no longer being held in check by a rational force. What should have been suppressed by a normally functioning brain bubbled to the surface and manifested itself into a terrible action. Unable to accept the fact that the men the soldiers thought were the enemy would be released, the soldiers killed them on the spot. The rational decision-making prowess that separates us from animals had collapsed into rage. In a war that depended on winning over the hearts and minds of the locals, a few sleep-deprived soldiers made the Americans appear like warlords who administered justice in the streets. Something had to change.

On the type of hot, humid day in Washington, D.C., that reminds residents that the city was built on a swamp, I picked up a small red Toyota from the airport and headed out on the highway. I was on my way to the Walter Reed Army Institute of Research, the Department of Defense's main facility dedicated to testing the limits of the human body. I inched along the Beltway and soon found myself driving down a small street in Silver Spring, Maryland, lined with strip malls and a kitchen supply store called Counter Intelligence. The red brick wall surrounding the military base soon popped into view. With gleaming glass towers, it looked more like the campus of a major corporation than a collection of barracks.

I was there to meet Thomas Balkin, a civilian scientist who has spent the last twenty-five years working for the military and who was now the chief of Walter Reed's Department of Behavioral Biology. A few weeks before I met with him, Balkin had traveled to Europe and made a presentation to top NATO brass on how to prepare soldiers for warfare in the twenty-first century. Through this presentation and others, Balkin was in the midst of a campaign to convince war planners that the success of future missions depended on recognizing the innate weaknesses of the human body that can impact the abilities of soldiers to make smart choices. The management of sleep and fatigue topped his list.

"A sleep lab isn't the most exciting thing to see in the world," Balkin said as he took me on a tour of the rooms where he has ushered in hundreds of participants of sleep deprivation studies. With government-issued blue couches, a small TV, and a shelf full of video games, the rooms looked like a suite in a college dorm. We made our way into an office so small that it couldn't hold more than four people at once. Balkin sat down behind his desk, and with a couple of clicks of his mouse, he opened a graph on his computer that showed a red line plummeting. "See this line?" he asked me, pointing with his left hand. "This represents the number of reported incidents between a soldier and a civilian in Iraq. Twenty percent of the men who are sleeping less than four hours a night have reported an altercation with a civilian. Follow it all the way across, and you'll see that only 4 percent of the men who are sleeping eight hours a night have."

Lack of sleep was an underlying cause of an exhausting and depressing cycle, Balkin explained. Grumpy, tired soldiers have less control over their emotions and are therefore more likely to get into a fight with civilians. Those civilians, in turn, are more likely to have a negative view of American forces and their presence in the country. Some vent by harassing U.S. forces, which of course leads to additional sleep-deprived soldiers. It was a cycle that captured the dual sins of being self-inflicted and preventable. Instead of helping to secure an area, pushing soldiers past the line of sleep deprivation was undermining the success of the mission.

Sleeplessness has always been a part of combat, but its effects are greater now because the military asks more of its soldiers than it did a generation ago. On the surface, this seems silly. Technology has allowed many of the jobs that were once performed by humans to be automated. But automation works only when the human operators—the men and women sitting in the chairs directing the show—make smart decisions again and again. Instead of overtaxing each soldier's body, the military now overworks a soldier's mind. Take, for instance, a new class of navy destroyers that reduces the number of sailors needed aboard from three hundred to less than one hundred. With only one third of the personnel, the new ships are able to execute a wider range of functions. Technology increases the human responsibility when measured in terms of the decisions each sailor has to make in order for the destroyer to perform. That means that the errors of one sleep-deprived sailor will reverberate throughout the whole ship because there is little in the way of backup. In places like Iraq and Afghanistan, meanwhile, troops have to make countless choices when trying to distinguish between civilians and potential terrorists. "Guys who are patrolling are constantly surveying the environment looking for signs of potential danger," Balkin says. "With sleep loss you are less likely to notice these things or notice them relatively slowly."

To solve the problem, Balkin realized that the military needed a way to measure and prepare for sleep as accurately as it does everything else. This is made more difficult by the fact that, unlike with alcohol in the bloodstream acting as a measure of drunkenness, there is no absolute biological marker of fatigue. Humans are notoriously bad at accurately estimating how many hours they slept the night before, making most sleep data collected from soldiers unreliable. Moreover, peer pressure and demands from their commanders would likely push troops into overreporting the number of hours they slept and working without taking a break, even if their mental performance was severely restricted. "If we could discover a marker for it, it would revolutionize our ability to manage sleepiness," he told me. "You can't manage what you can't measure."

Balkin turned to the next best thing: a program first developed by the air force for pilots, one of the few professions in the military that come with hard restrictions on the number of hours a person can work in one week. The premise of the sleep-tracking model was simple: the longer a person stays awake, the less competent he or she will be. Through research studies, the military knew that cognitive performance declines by about one-fourth for every twenty-four hours of time spent awake. Pilots who haven't met their mandated hours of rest are considered not competent to fly, a policy meant to prevent exhausted pilots from making decisions that endanger themselves or their aircraft. Balkin realized that he could take the scheduling tool used for pilots and apply it to every soldier in the field. After all, soldiers who weren't sleeping enough were picking fights with civilians in Iraq, which was endangering the success of the military's top mission. For it to work, each soldier would need to wear a wristwatch-sized sleep monitor, called a wrist activograph, at all times. The monitor recorded the body's movements each minute to determine whether the person wearing it was asleep or awake.

Balkin decided to test his theory that hours of sleep alone could predict a soldier's performance. He went to an officer-training school and asked a number of cadets to wear the sleep monitors for several days. Periodically, he would assemble them in a room and give them an exam. He then compared their test scores to the estimated amount of sleep recorded from the sleep-monitoring system. "Soldiers that consistently averaged the highest amounts of sleep obtained consistently high exam scores, whereas those that averaged low levels of sleep obtained inconsistent performances on the exams, with some doing quite well and others receiving failing or only marginally passing grades," he summarized in a research report. In the classroom, inconsistent test scores were nothing more than annoying and a sign that a cadet needed to hit the books. But in the field, inconsistent or poor decisions could cost lives.

The monitor, which is expected to become a standard part of a soldier's gear by the end of 2020, opens up a flood of data to military planners that can be used to predict performance. Suddenly, with a few clicks of a mouse, a commander will know how many hours each person in the unit has slept—and, by extension, what kind of decisions he or she will likely make. Performance on tasks that range from maintaining friendly relations with civilians to changing strategy in the middle of a battle can be modeled based on sleep time, a system allowing operations to be more efficient and soldiers less likely to make a mistake that can have far-reaching effects.

Balkin imagines a scenario where a commander, through the data obtained from each soldier's wrist, realizes that a unit's decision-making abilities will start to decline in a few hours because of a lack of sleep. Depending on the demands of the battlefield, the commander can order the unit either to take a nap before it leaves on patrol or to ingest a stimulant such as caffeinated gum. Fatigue, long the overlooked nemesis of military efficiency, can soon be regulated and quantified as easily as food rations or bullets. In one report, Balkin estimated that in future conflicts, the number of friendly-fire accidents will plunge toward zero, all on account of the increased decision-making abilities made possible by sleep.

The ability to track sleep allows the military to hone its most vital asset—each soldier's intellect. Rather than trying to run its enemies into the ground with exhaustion, the military will harness sleep to mount an organization that has the ability to make smart decisions consistently, a battlefield edge that can trump advances in technology. Friendly fire may become a thing of the past, as confusion and exhaustion no longer give the enemy unearned advantages or lead to unnecessary deaths. “Sleep has always been a weapon,” Balkin said to me in his office. Now, he said, the U.S. military will be able to control it unlike any other organization in history.

Sometimes, however, sleep can transform your body into a weapon in ways that you didn’t intend. That’s what happened to a man named Ken Parks, at least. On one night in the suburbs of Toronto, he unintentionally revolutionized our understanding of deep sleep and the brain’s ability to transition between previously unknown stages of consciousness. And his actions led to one of the most perplexing questions popping up in courts across the world: Does accidentally killing someone while you are sleepwalking make you a murderer, or a bystander to a terrible mistake over which you had no control?

8

Bumps in the Night

It was not a good night to be Ken Parks. As he tried to wedge all six feet five inches of his frame onto his living room couch, Parks couldn't have known that he was about to upend our understanding of what the brain can do. At that moment, all he could think of was how he was screwed. He was broke, out on bail, and sorely in need of a time machine that could undo the last twelve months of his life. His wife wouldn't even let him into their bedroom.

It wasn't always like this. Though he was a high school dropout, he had convinced his wife, Karen, the daughter of an engineer, to marry him, and together they had fashioned a normal middle-class lifestyle. They owned a home in the suburbs of

Toronto and had a five-month-old daughter. Then he discovered horse racing. All it took was watching a five-dollar bet turn into forty-five dollars and he was hooked, convinced he had a supernatural ability to pick winners. That reality proved otherwise didn't matter.

With his large size, Parks was easy to spot at the track, making bet after bet and quickly draining the family's bank account. Once he ran out of money, he reasoned that the only way to get it back was to double down. After loan sharks refused to extend him any more credit, Parks created fake purchase orders at his job and deposited the money into his own account. He embezzled thirty thousand dollars before someone caught onto him. Stuck in jail and charged with fraud, Parks called Karen and told her that not only had he lost his job but that the family was penniless, and all he had to show for it was a mound of worthless racing slips.

Learning that your husband promised your refrigerator to a loan shark so that he could bet on horses would be enough to send any woman to a divorce attorney, but Karen was resilient. She told him that unless he promised to stop gambling, she would leave and take their daughter with her. She also demanded that Ken go to her parent's house and ask for their help in sorting out the financial mess he had made. Without much support from his own parents, Parks worshipped Karen's family. Now, they would know that he was nothing more than a screwup. With that unpleasant conversation scheduled for the next day, Parks lay in the house that he could no longer afford and watched *Saturday Night Live* in an effort to relax enough to fall asleep after two restless days without it.

That was the last normal moment in Ken Parks's life. Sometime during the night, he got off of the couch, walked out the front door, and got into his car. He then drove fourteen miles on a busy highway and pulled up to his in-laws' townhouse. He parked, got a tire iron out of the trunk, and let himself into the house with his key.

A couple of hours earlier, Ken's father-in-law, Dennis, had laid down next to his wife, Barbara, and fallen asleep. He was suddenly woken up by the disturbing realization that a very heavy man had his strong hands around his throat. "Help me, Bobbie! Someone's trying to strangle me!" he choked out. He kicked his legs in panic and soon lost consciousness. When he came to, he had no idea what time it was, why he was lying face down on the bed in a pool of blood, or why there was a police officer in his bedroom. His wife's body lay in the bathroom a few feet away. She had been stabbed five times and beaten over the head with a tire iron.

Around the same time, a large blond man with a dazed look in his eyes and blood covering his body walked into a police station a few blocks away from the home. As soon as she saw him, the patrol sergeant at the desk called for immediate backup. The man didn't notice that his hands were cut so deeply that a pool of blood was collecting under him with every step he took. "I've just killed two people. My God, I've just killed two people," he told the officer. He looked down and, as if registering his mutilated body for the first time, screamed, "My hands!"

It was only after the officers hastily bandaged the man's hands and put him into an ambulance that they learned his name. He calmly told them that he was Ken Parks. When asked whom he had killed, Parks replied, "My mother- and father-in-law," not realizing that Dennis had survived. An officer then asked him how he had done it. Had he shot them? Stabbed them? Ken's head jumped at the word. "The knife's in my car," he gasped.

A bloody man had walked into a police station and confessed to killing two people and, for good measure, told investigators where to find the murder weapon. Few cases have seemed so easy to solve. But as the detectives started unraveling what happened that night, something didn't quite add up. Aside from the embarrassment at disclosing his gambling debts, Parks had no motive to kill his in-laws. He knew that neither he nor Karen were in line to collect payments from her parents' life insurance policy, so their deaths wouldn't solve his financial troubles. There were no signs that he had lost his temper in an argument after he asked for a loan. If he had gone over to their house intending to kill them, then why would he grab a tire iron from his trunk when an axe lay right next to it? Why would a killer plan an attack and then drive directly from the scene of the crime to a police station? And, most bothersome of all, why couldn't Parks remember anything? In one unnerving moment of the investigation, a detective walked into Parks's hospital room a few hours after he was admitted and began to interview him. Parks asked if his in-laws were dead. The detective said that one of them was. "Did I have anything to do with it?" Parks asked. The officer couldn't tell if Parks was delusional or the best actor he had ever seen.

Later, at the trial, one of the most esteemed doctors in his field would give a simple explanation for why a man would go out and murder one of his in-laws and nearly kill the other on an otherwise boring Saturday night: he was sleepwalking.

A checklist to determine whether someone is asleep would seem to be pretty short. Eyes closed, check. Slow rate of breathing, check. Little reaction to the surroundings, check. Maybe some light talking or kicking could be involved, but definitely not driving, and certainly not murdering. However, as Ken Parks unintentionally discovered, it is possible to break all of these rules and still be asleep. The brain, as we now know, never really shuts down during the night. Instead, parts that are responsible for different functions turn on and off at various points throughout the sleep cycle. It is a little like a car factory that runs twenty-four hours a day, with workers responsible for painting coming in at noon and the crew that puts in seats arriving to work at six. When something happens to alter those delicately timed cycles, strange things happen.

Sleepwalking is the best-known condition of what are called parasomnias, a broad group of problems that arise when one part of the brain shows up for work when it is not supposed to or misses its shift entirely. In most cases, the result is a person who is literally half asleep. When someone is sleepwalking, the parts of the brain that control movement and spatial awareness are awake, while the parts of the brain responsible for consciousness are still asleep. That means that sleepwalkers can have their eyes open and react to the events going on around them, but have no conscious thought or memory. Though parasomnias weren't fully understood as a class of disorders until the early 1980s, Shakespeare was eerily correct in his description of the sleepwalking Lady Macbeth. In one scene, she sleepwalks into the room where two men are talking. "You see, her eyes are open," one says when Lady Macbeth walks by. "Aye, but their sense is shut," the other replies.

Although the reasons are unclear, about one in five people will sleepwalk at least once in their lifetime, though most outgrow it by the time they are in middle school. Sleepwalking children are relatively mellow and lethargic, whereas adults who wander around at night tend to make quick movements, as if they are in a hurry to do something. Scientists can't yet account for the difference.

Sleepwalking isn't the only complex behavior that can happen in your sleep. People with parasomnias can do pretty much every basic human activity while sleeping, including talking, eating, driving, masturbating, and having sex. (Some people with sexsomnia, as it's called, are better lovers while they are asleep than when they are awake. As one doctor told me, "This condition is only a problem if the person sleeping next to you doesn't like it. No harm, no foul." The same doctor also said he believed a better term for the disorder would be *snore-gasm*.) The only difference is that they have no conscious awareness of what they are doing. It's as if their bodies have rebelled and decided to go about their business without the brain's input.

In the early 1980s, two doctors at the Minnesota Regional Sleep Disorders Center in Minneapolis began cataloging the number of patients who complained of injuring themselves or their bed partners in their sleep. As part of their investigation, Mark Mahowald and Carlos Schenck videotaped each patient sleeping at least one night in the hospital's lab. What they found was a window into a bizarre world. Older men with sweet dispositions turned into angry sailors in the middle of the night, repeatedly swearing and punching the wall next to them. Other patients suddenly sat up, stared intensely at the wall, and then dove headfirst into the nightstand. And at least one man sat at the foot of the bed, bellowing show tunes while fast asleep.

Parasomnias seemed to be a particularly male trait. The wives of some patients came in complaining that their sleeping husbands had put them in a headlock the night before or had tried to strangle them. Not surprisingly, these couples would often end up sleeping in different rooms. In a series of interviews conducted by the doctors, one woman said that her otherwise mellow husband would get out of bed during the middle of the night and crouch in the corner of the room and snarl like a wild animal. Another said that her sleeping husband repeatedly destroyed the furniture. "He broke so many lamps in our bedroom," she said. "You don't want to spend any money on lamps because you know they'll be flung across the room." Other patients with parasomnias said that they had flung themselves out of second-story windows while sleeping. This usually happened only once: after the first time, patients took to tying themselves to their beds at night out of fear that they might inadvertently commit suicide. Patients told the doctors tales of getting into a car and driving more than ten miles to a family member's home, of running down the street with dogs nipping at their heels, and of almost snapping someone's neck in their hands—all while sleeping. Out of these and other cases, Mahowald and Schenck were the first to identify and classify what are now known as violent parasomnias. As with sleepwalking, almost all are caused by partial arousals in the brain.

Like blond hair or high arches, parasomnias run in certain families. My own habit of sleepwalking, for instance, may be something I inherited. It wasn't until after I began working on this book that my father told me stories of his sleepwalking on the farm where he grew up, in Kansas. More than once, he told me, he woke up in his pajamas in the middle of a cornfield.

The Parks family tree was not the place to look for normal sleep patterns. Or dry sheets. The Parks men shared the embarrassing habit of wetting the bed, a condition that scientists tactfully call enuresis, well into their teens and twenties. Doctors blamed it on staying in deep sleep for too long. All would later sleepwalk as adults. In a bizarre variation of the midnight snack, Ken's grandfather would often sleepwalk into the kitchen and start frying eggs and onions on the stove, only to go to bed without eating. The family traits didn't skip Ken. When he was eleven, his grandmother caught him trying to climb out of a sixth-story window while he was sleepwalking.

Parasomnias, and sleepwalking especially, can be triggered

by sleep deprivation. As the brain struggles to make up for its lost sleep, it stays in the deeper stages of sleep for a prolonged time and doesn't always make a smooth transition to the next stage. Those rough handoffs lead to weird behavior. Ken, after spending two sleepless nights worrying about his marriage and the money he owed, was perfectly primed for a sleepwalking episode.

There was no question that Parks killed his mother-in-law. The only real question was whether he was sane on the night of the murder. The British, Canadian, and American legal systems all come from English Common Law and have wide areas of overlap. In each country, criminal law is based on the idea that in order to be found responsible for a crime, a person needs not only to have committed an act but also to demonstrate *mens rea*, or a guilty mind. This is how we separate accidents from crimes. If your brakes go out suddenly while you are driving on a busy street and you hit and kill a person, you won't be charged for murder, even though you were responsible for another's death (whether you should have known that your brakes were on the fritz is another matter). But if you purposely used your car as a weapon, then deadly action was paired with deadly intent.

The precedent of considering the mental state of the wrongdoer goes as far back as ancient Babylonia, where people who knowingly broke the law were put to death in much more gruesome ways than those who simply made a very bad mistake. In Greek mythology, Hercules was forgiven for killing his children during a rampage because he was put under a spell of madness by Hera, who was both Zeus's wife and sister and clearly dealing with some of her own issues. In 1843, the notion that a mental disorder nullifies criminal responsibility was codified in what became known as the M'Naghten rule, named after a case in which Daniel M'Naghten, a schizophrenic, shot and killed the secretary to the British prime minister while in the midst of a paranoid delusion. After he was found not guilty, public outcry that stretched as far up as Queen Victoria led the House of Lords to pass the first laws establishing the limits of the defense of insanity. Though it has been tweaked and modernized, the basic gist of the law still applies as written: "to establish a defense on the ground of insanity, it must be clearly proved that, at the time of the committing of the act, the party accused was laboring under such a defect of reason, from disease of mind, and not to know the nature and quality of the act he was doing; or if he did know it, that he did not know he was doing what was wrong."

Observers in the Parks trial thought that if Ken was indeed sleepwalking, he had a chance at acquittal under the definition of insanity. But simply pleading insanity doesn't mean that an accused person is given a free ticket home. Instead, many are sent to mental institutions for the rest of their lives and given no chance of parole. It can be just as bad as prison. Parks refused to claim that he was insane because he thought it meant he wouldn't see his daughter again.

His attorney was left to develop a novel defense. Sleepwalking, she said, wasn't a defect of the mind. It was simply a normal condition in which the body acted without any conscious input from the brain. Therefore, she reasoned, Parks couldn't be held responsible for something he never chose to do, and couldn't be deemed insane for a common and temporary state. In effect, she was asking for the jury to agree that Parks's mind was both not guilty and fully rational, even though his body committed a crime. It was the first time in Canadian history that a defendant claimed to be asleep when he was killing someone.

During the trial, Parks's attorney called Roger J. Broughton, an assistant professor of neurology at elite McGill University, to the stand. A few years earlier, Broughton had published an influential article in *Science* magazine in which he argued that contrary to Freudian reasoning, which viewed moving during sleep as an acting out of blocked emotional pain, disorders like sleepwalking or sleep talking were not a result of a person's mental state. The idea that sleepwalking reflected emotional turmoil was actually much older than Freud's theory. In a theatrical device that reflected the understanding of the time, Shakespeare implies that the guilt of murdering her husband's rival is what pushes Lady Macbeth into sleepwalking.

Broughton told the jury that Parks was most likely in deep sleep from the time he got off his couch to the time he walked into the police station. That would account for his lack of memory and for his apparent lack of motive. Broughton's theory was that Parks drove to his in-laws' house while acting out a dream, and that his series of movements would have been suppressed by hormones in a normally functioning brain if it wasn't for his genetic predisposition toward sleepwalking. Once there, his mother-in-law tried to wake him up, and that was when he went into a violent rage. No one knows why, Broughton told the jury, but sleepwalkers often react aggressively when confronted.

On cross-examination, the prosecutor asked how Parks was able to safely drive all fourteen miles of a route that included three stoplights if he truly was sleepwalking. Because his eyes were open, Broughton answered. Just as sleepwalking children can perform complex maneuvers such as walking down a flight of stairs without falling and hurting themselves, he argued, Parks was able to drive that night on a very familiar route without getting into an accident because he was essentially on autopilot.

Perhaps swayed by Broughton, or the fact that Karen testified in Ken's defense, the jury acquitted Parks of all charges after only a few hours of deliberation. Though it was obvious he had killed one person and came very close to killing another, the jury found that he did not do so voluntarily. Instead, Parks's deeds were classified under a new category—officially called a non-insane automatism—that allowed him to walk outside of the courtroom a free man.

After the verdict, frustrated prosecutors filed an appeal in hopes of preventing a surge of defendants claiming that they, too, had been asleep at the time of a crime. During the hearing, justices on the court puzzled over whether Parks's sleepwalking was so extreme that it could be termed a disease, like schizophrenia, and thus fit within the definition of insanity. But without any scientific evidence to support different degrees of sleepwalking, this argument fizzled away. Next, they discussed whether Parks was a ticking bomb, primed to kill again in his sleep. But again, there was nothing in the medical records to suggest that a person who has one extreme sleepwalking episode is likely to have another. In a somewhat exasperated decision, the court finally ruled, "This case is extremely troubling. The facts are so extreme that it stretches credulity to think that a person could perform all of those apparently deliberate acts over such an extended period of time without volition or consciousness. But the wisdom of the jury's verdict cannot be the subject of review in this court." It continued: "When asleep, no one reasons, remembers or understands. Medical experts do not understand exactly why those faculties do not function during sleep but it is accepted that they do not . . . If the respondent's acts were not proved to be voluntary, he was not guilty."

Nearly twenty years after Parks left the courtroom a free man, Michel Cramer Bornemann steps behind a lectern in a hotel banquet hall in central New Jersey. With a black shirt, red tie, pale skin, and black hair, he looks like he would be equally at home in a room full of Johnny Cash impersonators. A few hundred physicians and medical researchers have gathered here on a cold October morning to listen to him give the keynote address at the New Jersey Sleep Society's Annual Educational Symposium. Cramer Bornemann is an assistant professor of neurology at the University of Minnesota and has published several influential research studies, but that's only part of the reason why he is attracting a crowd. It is his side job that leads to calls from lawyers and law enforcement officials from around the world, to numerous requests to write articles for scientific journals, and to a recent symposium given in his honor at Cambridge University.

All of the attention comes from the fact that he is perhaps the world's only doctor devoted to investigating the strange habit of humans to commit crimes while they are asleep. In a way, he is the personification of the legal and medical worlds that Ken Parks created. Cramer Bornemann's life is spent looking at cases of sleepwalking, sleep driving, and sleep sex, all because he sees it as the next step toward reaching a full understanding of what it means to be conscious. He is a detective of the human mind, living in a world in which what Parks did was odd only by degrees. "What in essence we're doing is developing and defining a new field of sleep forensics," he tells the crowd.

It is a field with a greater demand than you might expect. Even before Ken Parks, a stream of defendants accused of violent crimes claimed to be sleepwalking, including at least two unrelated cases involving some unlucky person receiving an axe to the head. Few of them were successful. Albert Tirrell, the son of a wealthy shoe manufacturer in Boston, is thought to be the first person to cite sleepwalking as a defense and win. That came in 1846, well before science had any understanding of the nature of sleep or sleep disorders. Tirrell's attorney was able to convince the jury that his client slit the throat of a prostitute and burned down her brothel while in the throes of a nightmare. He wasn't so skilled at getting him out of the charge of adultery, however, and Tirrell ended up spending three years doing hard labor at a state penitentiary. Thirty years later, a Scottish man was acquitted of killing his young son while asleep, but he was freed only after he agreed to never sleep in the same room as someone else again. Sleepwalking cases popped up here and there in the following decades.

Then came the Ken Parks ruling. Within seven years after his acquittal, there were in Canada alone five well-known cases of defendants claiming to be sleepwalking when committing a crime. Around the world, cases of sleep violence were increasingly put before juries. In 2009, a fifty-nine-year old Welsh man was acquitted of murder after he claimed that he was sleeping when he strangled his wife of forty years in their motor home while they were on vacation. Like Parks, his attorney argued that he was not insane and thus did not belong in a mental institution. The judge and jury agreed. "You are a decent man and a devoted husband," the judge told him from the bench. "I strongly suspect that you may well be feeling a sense of guilt about what happened that night. In the eyes of the law, you bear no responsibility."

Surely some people who use sleepwalking as a defense are lying. But some are not. Those are the cases that Cramer Bornemann looks for. While he likes to help district attorneys identify liars who attempt to weasel out of their crimes by claiming to be asleep, he is most interested in the truth-tellers who give him a chance to document the strange capabilities of the sleeping body.

No one really knows how often a night's sleep leads to a crime scene, in part because of the nature of the things that sleepwalkers do. The brain isn't able to formulate a plan while awake and then carry it out while asleep, which means that there are no Bernie Madoffs or John Dillingers in the annals of sleep crime. Sleepwalking incidents that result in causing someone else pain tend to fall between two extremes, neither of them pleasant. In most cases, a person who acts violently in his sleep is a threat to whomever he or she shares a bed with. If a woman throws an elbow and breaks her boyfriend's nose while they are both asleep, the only authority involved will most likely be the doctor who bandages him up. Because there are no police reports or any other official records, there is no way to know how often these incidents of nocturnal violence happen unless they find their way to a courtroom. One study in the *Journal of Trauma: Injury, Infection, and Critical Care* cited twenty-nine cases of sleepwalking that resulted in an injury either to the sleepwalker or to those nearby. "There is a large tolerance from the family and even the medical community to episodes of somnambulism [sleepwalking], and it seems that the potential life-threatening risks associated with nocturnal wandering are not well understood," its authors noted, with a tone of bewilderment. This is often the case with REM sleep behavior disorder, a rare condition that mostly affects older men in which the brain doesn't paralyze the body during REM sleep like it is supposed to. The result is that patients act out their dreams. The wives of patients with this disorder often tell the sleep doctor stories of their husbands hitting them, and worse, while still asleep.

Cramer Bornemann investigates the other extreme of sleepwalking. Determining whether someone was asleep while committing violent acts ranging from child molestation to murder could mean the difference between freedom and the death penalty. That's not all. In a study published by the *Journal of Forensic Science* in 2003, Cramer Bornemann detailed cases of people falling from hotel rooftops, getting hit by cars after marching into traffic, and picking up loaded guns and shooting themselves—all while sleepwalking. The official term for this is parasomnia pseudo-suicide. Determining that these were accidents, and not intentional, can have profound emotional effects on the victim's surviving family members. Identifying sleepwalking as the cause can also trigger life insurance policies that don't cover suicide.

Cramer Bornemann didn't set out to be the Columbo of sleep crime. Not long after medical school, he was studying patients with amyotrophic lateral sclerosis, also known as Lou Gehrig's disease, at the University of Minnesota as part of an investigation commissioned by the National Institutes of Health. In a clinical trial, his team found that one of the first signs of respiratory problems associated with the disease surfaced during the later stages of non-REM sleep. Patients essentially had to work harder to breathe when they entered the deepest stages of sleep. Armed with this early clue, doctors could offer patients respiratory support before any larger symptoms arose. "From that, I became so enamored with sleep that I transitioned my career to focus solely on it," he told me.

He chose the best place to do it. His office door was just a few floors away from the offices of Mark Mahowald and Carlos Schenck, the doctors who identified parasomnias and helped make the University of Minnesota one of the world's foremost centers of sleep research. Based on the steady number of calls that the hospital's sleep lab received from prosecutors dealing with defendants claiming to have a committed a crime during sleep, Cramer Bornemann realized that violent parasomnias may be more prevalent than originally thought. Even if everyone he studied was lying, he reasoned, he would gain valuable insights into how to spot a fake. He approached the hospital's board, and soon afterward, Sleep Forensics Associates was recognized as a formal division in the neurology department. Cramer Bornemann has reviewed more than 130 cases since then. In almost all, he has looked for evidence of brain activity suggesting that an action was voluntary, following the same standard established by the Parks case.

Though it sounds like the sort of investigation that would have detectives bagging pillows for fingerprints, sleep forensics relies much more on witness testimony than physical evidence. Depending on what a defendant did and how he responded, Cramer Bornemann can extrapolate which parts of the brain were most likely working at a specific time. That can help him determine whether a person was sleepwalking or driven by another parasomnia. His goal is to determine how closely the subject's brain came to what we think of as awareness. "What we are starting to understand is that as we transition from wake to sleep, we can subdivide and identify various components of consciousness," he told me.

Several clues suggest which parts of a sleeping person's brain are within conscious control at a particular moment. The easiest of these to spot with the naked eye is muscle tone, which is completely absent in REM sleep. (Patients with REM sleep disorder are the exception to this rule, but their episodes are almost always over in fifteen minutes or less. Meanwhile, sleep-walkers, if left undisturbed, can move around for more than an hour.) The two areas of the brain that Cramer Bornemann pays the most attention to are the reticular activating system, which sits where the head meets the spinal cord, and the prefrontal cortex, the same area behind the forehead that is so important in thoughtful decision making. In some stages of deep sleep, both of these areas completely shut down. With them go the ability to repress impulses and, ominously, the ability to feel pain.

This can lead to strange forms of misfortune. One night in the middle of winter, a man in Minneapolis woke up in his bed with the uncomfortable realization that his sheets were wet. Embarrassed that he had apparently soiled the bed for the first time in his adult life, he pulled back the covers and found that his feet were black from frostbite. He immediately woke up his wife, who called an ambulance. As the paramedics were taking her husband outside on a stretcher, she noticed footsteps leading away from the front door and into the fresh snow. She followed them in a three-block circle around their house. In his sleep, her husband, barefoot on a night when the temperatures were in the low twenties, had gone for a walk, following the same route he took with the family dog. The dog had remained warm in its bed at home.

Cramer Bornemann uses this tale as an example of what he calls process fractionalization, a way of splitting up the stages of consciousness in sleep. That the sleepwalker was able to withstand the cold while walking barefoot through the snow suggested to Cramer Bornemann that his brain was not processing sensory input, one function of the brain's reticular activating system. It wasn't until the sleepwalker got back into bed that he felt the sensation of wetness from the melting snow on his feet. By that time, the man's brain was in a different stage of consciousness, allowing him to be aware of the wet bed and to wake up.

How does this relate to crime? Say a witness testifies that a barefoot man came after him with a baseball bat and stepped into a pile of broken glass but didn't react. Or say a man like Ken Parks cuts his hands to the bone and doesn't notice his injuries until he is spilling blood onto a police officer's desk. Because neither person seemed to feel pain, it's very unlikely that the parts of the brain that control consciousness were functioning at the time of the attacks. The body was taking actions independently of the mind, pushing these incidents into the realm of involuntary rather than criminal actions.

Prolonged sleepwalking incidents such as what occurred in the Parks case are rare, even in the new world of sleep crime. More often, Cramer Bornemann finds himself looking into cases of sexual assault, especially those brought on by alcohol. Almost half of the cases he studies involve a man sexually assaulting a woman or child after a night of heavy drinking. Cramer Bornemann argues that because binge drinking can lead to irrational behavior and slurred thought, the decision to down alcohol makes it impossible to tell whether a person was truly sleepwalking. Without any scientific backup, most defendants in these types of cases plead to a lesser charge if they can.

While science is on his side, Cramer Bornemann can't always say the same for the legal system. Judges and lawyers routinely scoff at the notion that a sane person could commit a complex, violent act involuntarily in his sleep. In San Diego a fisherman was found guilty of murder despite claiming that he stabbed his girlfriend to death while he was dreaming in his sleep about gutting a shark. At his sentencing, a judge with the sporty name of Gary Ferrari dismissed the defense completely, saying from the bench, "This whole business of committing a murder while sleepwalking . . . I think the best word is sophistry."

Cramer Bornemann thinks that the issue goes deeper than judges viewing sleepwalking as the latest implausible Twinkie defense. "We clearly conflict with the legal system," he tells me one day while on a break from reviewing a case. He has deep reservations about the adversarial nature of a trial, with each side presenting its own paid experts who offer self-serving opinions. The law isn't enough like science for him. "You have a medical expert on the prosecution and a medical expert on defense. To the jury they take equal weight. In science and in medicine, we don't have a bipartisan system. What we have in medicine and science is a consensus-driven peer-review process," he said. "If there's a question whether someone needs to go to surgery and it's a very risky procedure, what do we do? In medicine, we don't get two surgeons and sit down and talk. We get in an auditorium and we run a morbidity and mortality conference. We get the best and brightest in the institution and we debate and criticize all the different angles to a case. Science doesn't have to do with absolute truths. It has to do with likelihoods and probabilities." In a legal system that deals in competing absolutes, Cramer Bornemann wants room for shades of gray.

Nevertheless, he sometimes testifies as an expert supporting either the prosecution or the defense (donating the money that he is paid to the university hospital), because he wants to play a part in getting the legal system to standardize how it views sleep and consciousness. "These cases may be the first times the criminal system will have to deal with these definitions of consciousness. And they will have to deal with it. Otherwise you are going to have these erratic and unpredictable outcomes," he says. "The Parks case was a landmark case, but you still see very uneven renderings in courtrooms in how you handle a sleepwalking case. Sometimes you see full acquittals and sometimes in certain jurisdictions the attorney will say, 'Well, it's better to not try to go for a full acquittal' and try to angle for the lesser charge of institutionalization or a plea deal."

This often happens in violent cases involving children. A few months before we spoke, he received a call from an attorney who worked with the public defender's office in Alaska. The attorney's client was a man—almost all cases of sleep crime involve men, though Crammer Bornemann doesn't know why—who lived with his wife in a trailer park. The couple's sleep schedules were upended by the incessant cries of their colicky infant. After weeks of little sleep, they came to a compromise: one would stay up all night with the infant in the living room, while the other would get to have a relatively peaceful sleep in the couple's bedroom. They would switch places every night.

It wasn't a bad plan. The husband, however, had a history of sleepwalking. One night when it was his turn to stay awake, he fell asleep in the living room with the baby on his chest. During the night, he dreamed that a wild animal was attacking him and the only way to get away from the beast was by biting its head and throwing it off his body. When he woke up, he found his child, who appeared to be unhurt, underneath the coffee table. He picked him up and put him in his crib and then left for work. When his wife woke up a few hours later, she noticed that her son had bruises and bite marks on his head, and she immediately took him to the emergency room. Concerned by the unusual nature of the injuries, a nurse at the hospital called the police. Within a few hours, the father was arrested and charged with child abuse.

Cramer Bornemann agreed to review the case for the public defender's office. He studied the child's injuries and each parent's statement, and eventually came to the conclusion that the father could have accidentally hurt his son during a violent sleep episode. But he also realized that there was little chance his opinion would hold much weight in the legal system. "There are significant weights on the prosecution to prosecute any type of child abuse or any type of sexual assault. It's very unlikely that some prosecutor will say, 'Sleepwalking, that's a likely explanation,'" Cramer Bornemann said. In this case, the man agreed to plea a lesser charge before the trial.

Perhaps because he is developing a field focused on the outer limits of the brain's actions, Cramer Bornemann accepts that certainty can be elusive. "I can never really know what happened on that particular evening," he told me. "I'm different from other types of forensic investigators. I don't have DNA. I don't have tissue. I don't have any formal type of material evidence to provide confirmation. All I have is behavior patterns, and based on these patterns I can evaluate the brain state of a particular time and assess a likelihood, assuming people are being honest. In the court system we assume honesty or it's perjury. That's all we have to go on. "

Even then, the results can be unsatisfying. In January of 1997, Scott Falater, a forty-three-year-old software engineer at Motorola who was active in his church, went into his home office after dinner to continue working on a stressful project. He knew that if it failed, there would be a round of layoffs at the company. Like others who go into software engineering, Falater was more comfortable with numbers than with social interaction. "He's kind of nerdy," his daughter would later say. Work was taking its toll: Falater had been sleeping less than four hours a night for nearly a week straight, and had turned to caffeine pills to try to stay awake. Before he went to bed, his wife, Yarmila, asked him to fix the filter in the swimming pool in their backyard in suburban Phoenix. Around 9:30, he said good night to his family and went to bed.

About an hour later, a neighbor next door heard screams coming from the Falaters' house. He looked out of his window and saw what looked to be a woman's body lying in their backyard. He then saw Scott Falater approach the body while putting on heavy canvas gloves. Falater dragged the body to the pool and rolled it into the water. The neighbor ran to his phone and called the police. At the same moment, Falater was taking off his bloody clothes and putting them in a plastic bag. He then stashed the bag in the wheel well of the family Volvo, put a bandage on his hand, and changed back into his pajamas. That's how the police found him when they pulled up to the house moments later with their guns drawn. Falater was immediately handcuffed and arrested, and soon charged with first-degree murder. The coroner later concluded that he had stabbed his wife forty-four times with a hunting knife before holding her head underwater.

You can see where this is going. Falater claimed that he was sleepwalking at the time of the murder. His attorney called a number of witnesses who said he had no motive for the killing: Scott and Yarmila rarely fought, the family wasn't facing any financial problems, and there was no evidence that either had had an affair. A number of sleep researchers testified in support of Falater, including two who had testified in the trial of Ken Parks (Cramer Bornemann, who had yet to form his sleep crime agency at the time, was not part of either trial). The defense argued that the incident was so clearly an accident that the prosecution needed to prove Falater wasn't sleepwalking for there to be a case. The prosecution was having none of it. "They're so quick to want to make this thing a cause célèbre," one of the district attorneys told jurors. "I submit to you that [the credentials of the defendant's sleep researchers] are nothing but steps to their shrines of self-indulgence."

The jury deliberated for eight hours before it announced

its decision: guilty. Though Falater faced the death penalty, he was sentenced to life in prison. As of 2010, he was state inmate 148979 in a jail south of Tucson, a few miles from the Mexican border. He has worked as an aide in the prison library and as an academic tutor for the past several years, where he has earned excellent reviews for his behavior and continued to maintain his innocence.

Imagine, for a second, that both Parks and Falater are telling the truth. In our current legal system, which views consciousness as an all-or-nothing affair, the outcome of their sleepwalking could land them in a mental institution, on death row, or completely free. Those extreme options interested Deborah Denno, a professor of law at Fordham University in New York City who has the rare distinction of having earned both a JD and a PhD in criminology from the University of Pennsylvania. As a legal scholar, she has written several influential law-review articles arguing that penalties for certain types of crimes are too harsh. Four Supreme Court justices have cited her work in their decisions. Recently, she turned her attention to whether findings in fields like psychology and sociology can inform the criminal system. In an article published in a journal called *Behavioral Sciences and the Law,* she argued that the way the courts view sleep is outdated and in need of reform.

On one spring day, I traveled uptown to meet her in her office in Manhattan. Across the street, tourists were gathering for a performance of *South Pacific* at Lincoln Center. Not long before we met, she had returned from Tokyo, where she gave a lecture at a leading Japanese law school on consciousness and culpability in the American legal system. While we spoke, she sat perched on the edge of her chair. Her twelve-year-old shih tzu took up the rest. "The law is both too narrow and too broad when it gets to these issues of consciousness," she told me.

The legal system is built on the idea of uniformity: if an action is viewed as a crime in one jurisdiction, it is likely to be illegal in another. But, as Cramer Bornemann noted, that isn't happening with cases of sleep crimes. Denno found that some courts viewed sleepwalking as a voluntary act, meaning any actions arising from it were considered crimes for which a defendant should be punished. Others, however, viewed parasomnias as involuntary and let defendants go, often without a hearing, much less a trial. There was no middle ground between incarceration and freedom, and no record of how often police officers dealt with cases of possible sleepwalking. Part of the reason for that, she says, is that the criminal code hasn't been thoroughly updated since the 1950s, during a time when Freudian interpretations of consciousness still held considerable sway and the scientific understanding of sleep was rudimentary. Even now, the law sees all actions as either voluntary or involuntary. In her speech to the Japanese law students, Denno argued that this "is where the real injustice to a defendant lies."

She thinks the criminal code needs a third choice: semi-voluntary. "That may sound strange coming from me, as a person who argues for lesser penalties, but we should have a choice that would net more people into the system," she told me. As an example of what she means, she brings up Ken Parks. "I don't think that he should have been acquitted," she said. "There were aspects of his background that suggest that there should be another choice and a way to capture him into the criminal justice system. You may want to tell someone like Parks, 'If for the next year you take your medication and keep out of trouble, then we won't prosecute this crime.' "

Under Denno's system, there would be a record of every incident in which sleepwalking turns violent. If a person doesn't take steps to keep his or her condition in check—such as by taking clonazepam, a muscle relaxant that helps elderly men with REM sleep behavior disorder—then, Denno argues, that person should be held accountable for anything that happens while he or she is sleepwalking. In this system, sleepwalkers would be viewed in much the same way as loaded weapons: those who don't handle themselves responsibly can be found guilty of criminal neglect. The change would go a long way toward making sleep cases less of an anomaly and would help develop a fair and dependable standard, she says. It would also pull out of the shadows cases of sleepwalkers committing violence and give researchers something sorely missing in the world of sleep crimes: data. "If someone asked me how many sleepwalkers get through criminal justice system, no one would be able to answer that question," Denno told me. "Only if there was a murder would the courts ever see them. Otherwise, they are screened out of the system."

Until there is a system to track how often sleepwalkers commit crimes, we are left with the notion that there may be more cases than scientists or lawyers realize. If you are the sort of person who likes to wager, there are pretty good odds that the person sleepwalking in your home won't do anything more dangerous in the middle of the night than run into a wall. But infinitesimal chances sometimes happen.

For Ken Parks, life has gone on without any more incidents of sleepwalking leading to a courtroom. But it hasn't been easy. He and Karen divorced not long after he was acquitted. Since then, he has largely stayed out of the public eye, and perhaps having had enough of his time in the media, he frustrated all of my attempts to find him in Toronto. He is still living there, however, and has five children, if the local newspaper is to be believed. In 2006, he ran for a seat on the local school board. It didn't go over well. "Sleepwalking perhaps [could be forgiven], a medical thing, but not the embezzlement," one local man said when telling a reporter why he wouldn't vote for Parks.

Proving that there are some things in life that you can never live down, Ken came in last.

9

Game Time

Imagine, for a second, that you are in a casino in Las Vegas. You have been in town for a few days, spent too much and slept too little, and recently found out the hard way that you are not as good at poker as you had thought. Now would be a good time to find a wager where the odds, like a tipsy bartender, are tilted in your favor.

Here's a hint: find your way to the sports betting parlor. This is the dark, cavernous room usually filled with men huddled over small monitors and wearing looks of such intense concentration that you would think they were directing the launch of the space shuttle or planning a land-based assault against a rogue dictator. What they are actually doing is getting drunk and hoping to make money while watching wealthy athletes play games.

That's not to judge, of course. Betting on sports is popular because the games operate outside the realm of cruel probability. In blackjack, for instance, the dealer sliding you a 10 means that the odds that the next card will not be a 10 just went up. Yet in basketball, watching LeBron James dunk over an opposing center doesn't impact what will happen the next time his team gets the ball, or even the final outcome of the game. LeBron could always twist his ankle three minutes later, miss the rest of his shots, or not get the ball passed to him again because he offended his teammates during a time-out.

What makes betting on sports possible is the point spread. This is the bookmaker's attempt to impose order on the randomness of sports, and it is pretty much what it sounds like: given all of the known information about the game, it predicts that Team A is expected to beat Team B by a certain number of points. When you place a bet, you decide how you think that forecast is wrong. If you bet that Team A will score more points than predicted and you are right, you will make a little money; if you wager that Team B will pull off an upset and are right, you will make even more. If you're in luck, the next *Monday Night Football* game listed on the board will be a matchup pitting a team from the West Coast against a team from the East Coast. For most gamblers, deciding which team to bet on comes down to factors such as hometown loyalty, the trend of the last few games, or which team is on the road. But there's a much easier way to beat the odds: just put money on the team from the West Coast.

Betting tips wouldn't seem to fit in a book about sleep. Yet sleep is the most obvious part of a cycle that affects almost all living things, from quarterbacks in the National Football League to bacteria in the locker-room shower. Living organisms have an innate sense of the length of a day, which isn't all that surprising considering that, for the long history of life on earth, the sun has pretty much run the show. Plants rely on sunlight for energy, while animals have settled into niches based on the span of hours they considered the optimal time to be awake. As a result, somewhere inside the cells of most living things is what amounts to a fairly accurate twenty-four-hour clock, known as the circadian rhythm, which tells an organism when it is time to perform an important activity and when it is time to rest.

The first person to recognize this rhythm was an eighteenth-century French astronomer named Jean-Jacques d'Ortous de Mairan. Staring at his garden in 1729, he was struck by the fact that his plants extended their leaves in the daytime and then pulled them back at night. He assumed that it was based on sunlight, and he set up a simple experiment to test his theory. He took a bunch of plants down to his wine cellar, where there was no variation of light or temperature between day or night, and recorded the movements of their leaves. Sure enough, the plants continued to open in the morning and close in the evening, even though there was no daylight prodding them. De Mairan realized that the plants were anticipating sunlight rather than responding to it. They had an ingrained sense of when the day started, and they didn't need light to clue them in.

The human body is more like de Mairan's plants than you might expect. The circadian rhythm alters our body temperature and overall alertness level based on the time of day, a process that will make the body continue to stick to the sun's schedule even if it is placed in the equivalent of a wine cellar overnight. Without any help from coffee, most of us tend to perk up around nine o'clock in the morning and stay that way until around two in the afternoon, which is when we start thinking about a nap. Around six in the evening, the body gets another shot of energy that keeps us going until about ten at night. After that, our body temperature starts to fall rapidly, and we get sleepy if we don't turn to coffee or another form of caffeine. Evolutionary biologists don't know exactly why the body operates on this sort of split rhythm, but the best guess is that the early-evening pick-me-up was advantageous to early humans who needed energy to make a fire or find their way back home after a long day of foraging for food.

All of which brings us back to betting on football games. In the middle of the 1990s, a few sleep researchers at Stanford University decided to test a theory. Studies had shown that strength, flexibility, and reaction times surge in the early evening when the circadian rhythm is pulling the body out of the post-lunch funk. Given that subtle effect on an athlete's abilities, it stood to reason that a person at the peak of this alertness cycle would have an unseen advantage over someone whose internal clock thought it was time for bed. What the researchers needed to test their idea was a contest that not only pitted people of similar abilities at different stages of the circadian cycle against each other, but also a data set long enough to show reliable patterns.

They found it in the professional football games played on Monday night, some of the premier events in the NFL. Monday night games always start at 8:30 p.m. Eastern Standard Time, regardless of which teams are playing or which team has to travel. For the league, this guarantees the maximum number of viewers. Diehard football fans on the East Coast will stay up until past midnight if they have to, while sports fans on the West Coast can turn on the TV and watch the game while eating dinner after work.

The scheduling of *Monday Night Football* games presents a unique circadian problem, especially if a team from the West Coast is playing a team from the East Coast. Players on the West Coast team are playing at their equivalent of 5:30 p.m., no matter if the game is in Seattle or Miami. Players on the team from the East Coast, meanwhile, are three hours ahead in their own circadian cycles. In nature, this sort of mismatch couldn't happen. It was only in the last sixty years or so that we've developed a way to travel so quickly across time zones that our internal clocks are no longer in sync with the daylight around us. Fitting its cause, we call this condition jet lag.

Without knowing it, athletes on teams from the East Coast are playing at a disadvantage. Because of the circadian rhythm, which they can't control, their bodies are past their natural performance peaks before the first quarter ends. By the fourth quarter, the team from the East Coast will be competing close to its equivalent of midnight. Their bodies will be subtly preparing for sleep by taking steps such as lowering the body temperature, slowing the reaction time, and increasing the amount of melatonin in their bloodstream. Athletes on the team from the West Coast, meanwhile, are still competing in the prime time of their circadian cycle.

Every human body, ranging from a professional athlete to a suburban dad, will experience small declines in physical ability and mental agility the longer it fights against the circadian rhythm. In the modern NFL, this has a significant impact because teams in the league are more evenly matched than those in the other major sports, and anything that alters a single player's ability has an outsized effect on the outcome of the game. What's more, there is little that an East Coast team can do about the circadian disadvantage. The schedule gives coaches few chances to adapt to the time difference. Teams traveling on the road typically fly in the night before the game, and East Coast teams playing at home rarely attempt to put their body clocks on Pacific Standard Time. Coaches instead tell their players not to try to adjust to the time differences, preferring that they keep up with their normal sleep patterns for consistency.

The Stanford researchers dug through twenty-five years of Monday night NFL games and flagged every time a West Coast team played an East Coast team. Then, in an inspired move, they compared the final scores for each game with the point spread developed by bookmakers in Vegas. The results were stunning. The West Coast teams dominated their East Coast opponents no matter where they played. A West Coast team won 63 percent of the time, by an average of two touchdowns. The games were much closer when an East Coast team won, with an average margin of victory of only nine points. By picking the West Coast team every time, someone would have beaten the point spread 70 percent of the time. For gamblers in Las Vegas, the matchup was as good as found money.

In a test to ensure that their findings weren't the result of West Coast teams simply being better during those years, the researchers expanded their scope and looked at every *Monday Night Football* game played during that twenty-five-year time span. They found that the overall winning percentages for West Coast and East Coast teams were essentially even when the teams were not playing a game against an opponent from the other coast. Nor were the results a reflection of home-field advantage. When an East Coast team traveled to another destination within its same time zone, it won 45 percent of the time. But if a team from the East Coast played somewhere in the Pacific time zone, its winning percentage shrunk to only 29 percent.

Later studies found that circadian patterns influenced the outcome of other sporting events as well. In the late 1990s, Leonard Kass watched the University of Maine women's basketball team lose a game that they were favored to win in the National Collegiate Athletic Association tournament. The team had traveled to the West Coast for the game. "They just looked like they were out of phase," he said.

Kass, a big fan of Maine women's basketball, had more than a passing interest in the loss. A neuroscientist at the university who studies circadian rhythms, he decided to test how often a matchup of college teams on different circadian schedules resulted in an upset. He looked at several years of data from the men's national basketball tournament and found that higher-ranked teams that had to travel across the country were at a distinct disadvantage. "The kiss of death is shifting three time zones," he said. Teams that flew to the opposite coast were twice as likely to be beaten by a lower-ranked opponent in the tournament's first round. Circadian schedules trumped natural ability.

The circadian advantage—or disadvantage, depending on your perspective—popped up in studies of figure skaters, rowers, golfers, baseball players, swimmers, and divers. Everywhere you turned, there was evidence of the body's hidden rhythms at work. One study found that in sports as varied as running, weightlifting, and swimming, athletes competing when their bodies experienced the second boost of circadian energy were more likely to break a world record. Long jumpers, for instance, launched themselves nearly 4 percent farther when the body was at its circadian peak. But the circadian rhythm cut both ways. Athletes competing when their circadian rhythm corresponded to the so-called sleep gates—those times in the early afternoon or late nights when it's easy for most people to fall asleep—consistently performed a little worse than normal, even if the slowdown wasn't obvious to them.

Thanks to technology and medicine, every athlete at the college level and above is already close to his or her limit of sheer physical strength. In matchups at this elite stage, the tiniest advantage can mean the difference between winning and losing. In sports that require trials, such as gymnastics and figure skating, a disadvantage based on the time an early heat or qualifying round takes place can translate into a failure to get into the finals. Soon after these studies were published, athletic trainers became interested in the little-known effects of scheduling and sleep on physical performance. Here, they realized, was an unseen aspect of competition that could give them one of the few edges left in sports. Manipulating sleep and the circadian rhythm could be the last, untapped method of outflanking an opponent.

Charles Samuels is a plainspoken Canadian physician who seems more at home yelling in the stands at a hockey game than sitting behind a computer in a small office. Until 2005, his research focused on the health effects of working overnight shifts shown by police officers in Calgary, a city best described as the Canadian Dallas. That year, however, he started thinking about how the performance cycles that affected cops would influence the body of, say, a speed skater. "I'm interested in how sleep affects high-performing individuals," he told me. "The sports thing started off as a hobby. I didn't know that it was going to turn into something really quite demanding."

It was fortunate timing. Canada around that time had decided that when it came to sports, it was tired of being the nation of nice. The 2010 Winter Olympics were set to take place in Vancouver in a few years, and the Canadian government made it a priority to win more medals than any other nation. It set up a $6 million fund called Own the Podium for research and development, spending money on things like studying the aerodynamics of bobsleds in wind tunnels. Samuels submitted an application to come up with ways to harness the circadian rhythm. He had a two-pronged approach. First, Samuels needed to find a way for an athlete to quickly overcome jet lag after traveling across multiple time zones. Then, he wanted to manipulate the circadian dips that would affect an athlete's performance no matter when or where an event took place. The challenge was to master time and space in such a way that the body of, say, a Canadian sprinter competing in Beijing at 2:30 p.m. would think that it was really 6:00 p.m. and time for a boost of energy rather than a nap. Done correctly, mastering the body's rhythms could provide a slight advantage over one's rivals.

To do so, Samuels drew on research that went as far back as 1662, when French philosopher René Descartes correctly noted that the tiny pineal gland in the brain reacted to light hitting the eyes. He called this gland "the seat of the soul" because he believed that it was responsible for thought and the movements of the body. In the 1950s, researchers at Yale University discovered that one of the chief functions of the pineal gland is the production of a hormone called melatonin, which is released into the bloodstream at night. Like a parent singing nursery rhymes to a toddler, melatonin coaxes the body to fall asleep. High levels remain in the bloodstream into the morning. The discovery of this hormone helped scientists identify the gears behind the biological clock: a small cluster of cells deep behind the eyes called the suprachiasmatic nucleus (known as the SCN to biologists). These cells let the pineal gland know when the eyes pick up bright light. When it's dark for a while, the pineal gland assumes it is bedtime and sends melatonin throughout the body to let the organs know that it is time to close up shop.

But here's the thing: the pineal gland can be tricked pretty easily. For all of its wonders, the human body is still a reflection of a world where the only source of bright light was the sun. Abundant white light—especially white light with a slight blue tint that mimics the sky on a clear day—can fool the pineal gland into thinking that the sun is still up. That is why watching television or working on a laptop late at night can make it harder to fall asleep. The SCN registers the light from the TV and tells the pineal gland that it is a little bit of sunlight because that is all the brain is built to understand. With enough bright light over the right amount of time, the pineal gland becomes something of a reverse snooze button, holding off on releasing melatonin because it thinks that it is still daytime.

Jet lag is debilitating because, light-wise, the body has no idea what is going on. You can feel this yourself the next time you travel across time zones. Let's say that you are flying overnight from New York to Paris. When you land at eight o'clock in the morning Paris time, your body thinks that it is two o'clock in the morning, which it is back in the States. Walking out onto a Parisian street bathed in morning sunlight confuses the body even more. While the reasoning part of your brain grasps the concept of a time difference, your pineal gland only understands light. Because the human body wasn't designed to hop across an ocean overnight, this gland reacts to sunlight coming at what it thinks is the middle of the night and considers it a sign of an abnormally long day. In effect, the body becomes convinced that it is still the afternoon of the day before and takes steps to rewind the circadian rhythm back an hour or two. Now you have a bigger problem: your body isn't just six hours off of Paris time, it is eight hours off. And as anyone who has tried to fall asleep early for a few nights before a flight east across time zones knows, attempting to prepare the body for a time change is often an exercise in frustration. Many travelers simply can't fall asleep at seven in the evening, no matter how exhausted they are during the day, because of the circadian rhythm's early-evening shot of energy.

Samuels's plan for the Canadian Olympic team came down to a hyper-awareness of light and its effects on sleep. He created a plan where athletes would begin to shift their exposure to light before they traveled, starting one day for each time zone that separated them from the place of competition. This would allow them to adjust to the time zone quickly once they got there and, just as important, to significantly improve their sleep the night immediately before an event. If this is done correctly, a Canadian athlete would have a natural edge. "The research into sleep has been done, but no one cares about it," Samuels told me. "The U.S. cares even less. In the U.S. you'll find it very difficult on a global scale to get coaches and trainers interested in sleep to the same degree that we are in Canada."

He brought up a chart on his computer. "As I sit here talking with you, the lead of the BMX national team just sent me the upcoming world championship schedule. Soon, we will send her a team travel plan back with everything they need to do from the minute they get on the plane to the minute that they compete," he said.

Samuels believes that bending an athlete's circadian rhythm in preparation for competition will eventually usher in a new era of performance and sports training. "Ten years ago I would have been very reticent to say that light did this or that, but now we know that light improves alertness," he said. "I was around in the sixties when Gatorade was invented, and this seems like it has the same potential to affect performance."

Thanks to Samuels, the Canadian Alpine Ski Team travels with fifteen to twenty light boxes, a type of oversized flat lamp that simulates natural sunlight. Once at the site of a competition, the athletes will eat breakfast while sitting in front of one of them. The light stabilizes the circadian rhythm and improves alertness. Many will also spend ten minutes in front of a light box immediately before a race, especially if they are competing when their circadian rhythm normally dips.

Theoretically, this would improve their performance. I asked Samuels how one knows that the light strategy has worked in the real world.

"You don't," he replied simply. "My job isn't to sit here and take credit for wins because I worked with the teams. I never do that. These are athletes and they win and lose on their own."

Light improves the chances that athletes will compete at their peak no matter where or when the event takes place, he said. It is a form of sports training that isn't about charting lost body fat, lifting heavier weights, or ingesting the latest supplement. The focus is instead on harnessing the body's subtle rhythms so that an athlete doesn't walk onto a playing field with an unseen handicap. Beyond that, anything can happen. In all sports, athletes competing at their maximum ability sometimes lose.

But light is only part of how the circadian rhythm dictates athletic performance. The other part is actually getting athletes to sleep, which can be tougher than it sounds when they are constantly on the road. And perhaps no sport punishes athletes' sleep schedules over such a long time as professional baseball.

Fernando Montes has spent all of his adult life testing the limits of the human body. Unlike researchers in the military, he is not interested in the body's ability to withstand extremes of hot or cold or how long someone can go without food. Montes's interest in the body is limited to what makes it possible for one person to throw a baseball faster than another one.

Not long after he graduated from college, he found himself working on the strength and conditioning crew for Stanford's football team. After the team won its 1993 bowl game, Montes was offered a job as the strength coach for the pitching staff of the Cleveland Indians. The biggest change in jumping from football to baseball was the amount of time between one game and the next. The increased frequency of games radically altered Montes's concepts of strength and endurance. In football, his job was to shape brute strength, crafting a team full of athletes strong enough to bring down or run over the guy on the other team. But in baseball, sheer physical size wasn't as important as consistency. His new challenge was to train a pitcher's body to be able to hurl a baseball at speeds often over ninety miles per hour, more than a hundred times a game—and then do it again four days later. Added to this was a grueling 162-game, six-month schedule that would often include two straight weeks crisscrossing the country. Making the playoffs or the World Series could add another month and a half to the schedule. "In football the whole environment of recovery is not really truly understood because it doesn't need to be," Montes told me. "You have more days to recover because you're not playing everyday. It's a big difference from baseball, basketball, and hockey. Baseball is the worst because you're playing every night."

Schedules that feature six games in seven days make fatigue an unspoken part of baseball's culture. For a long time, players relied on amphetamines for a game-time boost of energy. Pills became a part of baseball in the wake of World War II, when returning veterans facing a double header turned to the same drugs the military gave them for use during combat. When the drugs were banned before the 2006 season, players said that their absence would be noticeable immediately. "It's going to have a lot bigger effect on the game than steroid testing," Chipper Jones, the Atlanta Braves' All-Star third baseman, said at the time. "It's more rampant than steroids. . . . I think the fringe players will be weeded out."

For most coaches, the long season boils down to a damned-if-you-do, damned-if-you-don't dilemma. Resting a top player so that he can perform at a high level late in the season comes with the risk that the absence will result in a loss today. Recovery was something that sports trainers in the United States concerned themselves with on a superficial level by giving athletes ice packs and massages, but Montes saw that sleep was rarely a part of the conversation. This seemed crazy. A pitcher's body won't recover from the trauma of the last game in time for his next start without sleep, for a number of reasons. Some of them are physical. Sleep, for instance, is the time when the body sends growth hormones to repair damaged muscles. But pitching is unique because it isn't only about muscle. Knowing the tendencies of your opponent—whether he will bite at a high curveball, how often he swings on the first pitch—is half of the battle. Without enough sleep, a pitcher might lose the ability to learn and analyze information that is vital to his success. A pitcher who isn't getting enough sleep has already lost the mental battle that he fights every time he is on the mound.

If carried out correctly, however, a schedule that emphasizes sleep could result in players who consistently play well. An athlete who can recover quickly creates an advantage that increases exponentially as the schedule gets longer, because it allows the best players to play in their top condition during the maximum number of games against progressively weaker rivals. If he could train players' bodies to regularly perform at their peaks despite the schedule and without banned drugs, Montes would find one of the holy grails of baseball. But one of his first challenges was to find an answer to something that bothered him from the minute he stepped onto a baseball diamond.

Pitchers, unlike other players on the field, aren't judged

on their foot speed, body strength, or any other standard test of physical ability. What makes a good pitcher isn't the ability to throw a ball faster than anyone else. It is the ability to throw strikes that batters can't hit. The result is a startling mix of body shapes. Yankee's All-Star pitcher CC Sabathia, for instance, is known as one of the most durable pitchers in the league because of the high number of complete games he finishes each year. He accomplished these feats while weighing close to three hundred pounds. In 2011, he surprised sports writers by showing up at spring training twenty pounds lighter, but downplayed the rumor that he was on a more stringent workout routine. "I stopped eating Cap'n Crunch every day . . . I used to eat that stuff by the box," he said. Sabathia, one of the top players in the league, was also one of the heaviest, and yet other top pitchers were tall and lanky. Age, too, seemed to have little effect on a pitcher's effectiveness. The Boston Red Sox's Tim Wakefield was forty-five at the start of the 2011 season, and already held the record for the oldest pitcher to ever play for the team. And yet he remained in the starting rotation on the strength of a knuckleball that rarely topped speeds of seventy miles per hour.

Montes wanted to find out if there was a measure that could help him evaluate his pitchers on a common scale. "In baseball, they keep stats up the wazoo," he told me. "So one of the first things I asked was, how can you tell me that a pitcher is in shape? Still to this day there isn't an answer, and as far as I know it was never studied by anyone." He turned to conversations he once had with sports trainers from the old Soviet Union who said that their tradition of emphasizing recovery time between events was an overlooked aspect of their success. Sleep obviously affected performance, and yet that didn't register within a sports world that often considers rest the sign of a soft athlete. Football's so-called Hell Weeks, for instance, are notorious for players' having to participate in full-contact drills twice a day for five days with little time for sleep or recovery. Yet this measure of toughness is deceiving. Surviving a Hell Week doesn't mean that a player will maintain his strength over the course of a full season. An athlete who can recover faster between games, on the other hand, has a clear advantage over competitors still sore from the last contest.

By the time Montes moved on from Cleveland to become the head strength and conditioning coach for the Texas Rangers, he had crafted a plan unique among professional sports trainers. From here on out, he decreed, his players would sleep. It was easier said than done. There was no controlling the travel schedule and the painful routine of arriving at hotels at three o'clock in the morning. So Montes set about searching for areas that he could control. He rounded up his pitchers and made each one record what time he went to bed, what time he woke up, and the quality of each night's sleep on a five-point scale. Hoping to speed the transition of each player's circadian rhythm to the time zone where the team was competing, Montes told pitchers to leave their curtains open when they went to sleep in their hotel rooms so that they would wake up with sunlight in their eyes. When one reliever seemed to be especially sleepy during home games, Montes cornered him. "Listen, I understand that on the road you like to go out at night. But at home, what's the problem?" he asked. The pitcher replied that he had young children. Sleeping at home meant giving his wife a break from handling them on her own.

Montes found himself armed with a rare accounting of his players' bodies over twenty-four hours. But it wasn't enough. He then decided that the pitching staff would report to the ballpark hours before everyone else. Pitchers arrived at the ballpark early, only to be told exactly when to take a nap and for how long based on the quantity and quality of their sleep over the past week. Like weight lifting, sleep became another part of training that required precision to be effective. "We had to teach them how to take a proper nap," Montes told me.

Proper naps were twenty minutes long, though each player was allotted thirty minutes to give him time to fall asleep. To make sure napping was easy, Montes set up an iPod to play what he describes as "relaxing meditation music" in a dark room. He made sure that a player's hands and feet were covered with a blanket, and lectured each man about the importance of keeping warm while asleep. If the head coach wanted to find one of his pitchers in the afternoon before a night game, Montes would tell him that he would have to wait until the naptime was over. Few knew it at the time, but the starting pitchers of the Texas Rangers that season often prepared for games in a dark room in the bowels of the stadium, riding out the dip in their circadian rhythms with a nap.

Baseball players, as a rule, are suspicious of experimenting with their training patterns. But after a week or so of the new sleeping routine, each player told Montes that he felt stronger and more energetic during games. Montes didn't want to cause friction with other coaches on the team who scoffed at the idea that napping could craft a stronger ballplayer, so he asked that his pitchers keep their new schedules quiet. It soon spread anyway. "One thing about baseball—and it doesn't matter if you're a pitcher or a position player—if you're successful, everyone wants to copy it," Montes said.

After one extra-innings game that went late into the night, the Rangers were packing up in the visitors' locker room in Kansas City. A plane would take them to Minneapolis that night. They wouldn't get into their hotel rooms until five the next morning. After less than ten hours at the hotel, team buses would arrive to shuttle them to the Metrodome for that night's game against the Twins. Montes went from player to player, recommending that they sleep with their blinds open and plan on getting to the ballpark early the next day to take a nap. It was an experiment to see whether his techniques could make a difference beyond his small circle of pitchers.

The Rangers essentially fielded two teams that night against the Twins. Players who didn't nap were out of sequence, missing what should have been easy defensive plays and struggling to connect at the plate. Those who arrived early at the ballpark to get extra sleep, meanwhile, performed about as well as they did any other night, and they demonstrated few of the side effects of the long night of travel and the sleep deprivation that had accumulated from the grueling road trip. The recovery plan wasn't enough to change the outcome of the ballgame—the Rangers lost that night—but the score was certainly closer than it would have been otherwise. The circadian rhythm wasn't conquered, but it was tamed. For the rest of the season, Montes's napping room was crowded.

It is all well and good that athletes can throw baseballs faster and speed down mountains quicker because they mastered the circadian rhythm, you might be saying to yourself. But how does this relate to people whose lives don't involve ball fields or coaches? The answer lies in one group that often spends several years of their lives in a constant state of sleep-deprived jet lag. They most likely live in your town, and may even sleep down the hallway from you. They go by a name coined less than eighty years ago: teenagers.

Edina is a wealthy suburb that sits less than ten miles outside of Minneapolis. Corporate executives and white-collar workers choose to live there in large part because of the quality of its public schools. It didn't seem like a place that would spark a radical change in education that still reverberates in school districts across the country. In the early 1990s, one of Edina's school board members attended a medical conference. There, he listened in rapt attention to a sleep researcher describe how teenagers' circadian rhythms differ from their parents' and siblings'.

Biology's cruel joke goes something like this: As a teenage body goes through puberty, its circadian rhythm essentially shifts three hours backward. Suddenly, going to bed at nine or ten o'clock at night isn't just a drag, but close to a biological impossibility. Studies of teenagers around the globe have

found that adolescent brains do not start releasing melatonin until around eleven o'clock at night and keep pumping out the hormone well past sunrise. Adults, meanwhile, have little-to-no melatonin in their bodies when they wake up. With all that melatonin surging through their bloodstream, teenagers who are forced to be awake before eight in the morning are often barely alert and want nothing more than to give in to their body's demands and fall back asleep. Because of the shift in their circadian rhythm, asking a teenager to perform well in a classroom during the early morning is like asking him or her to fly across the country and instantly adjust to the new time zone—and then do the same thing every night, for four years. If professional football players had to do that, they would be lucky to win one game.

The teenage circadian rhythm has become a problem only in the last hundred years. Teenagers before then were typically seen as young adults who worked to support the family, whether on the farm or in a trade apprenticeship if they lived in a city, and were given more control over their schedule. In 1900, only 8 percent of eighteen-year-olds had a high school diploma. By 1940, the proportion of high school graduates climbed to 30 percent, and by 1960 almost 70 percent U.S. teenagers finished high school. Although the quality of public education has vastly improved over that time, schools haven't been as kind to the teenage body. Teenagers in the past were expected to spend part of their days in the classroom and then either work at an after-school job or complete a round of chores on the farm. To fit in time for both, the school day started as early as 7:00 a.m.

This early start time remained constant despite sweeping cultural changes over the successive decades, including a sharp reduction in the percentage of young adults who work at an after-school job. Band practice, sports teams, drama club, and other activities that add to a college application have taken the place of paid employment for many teenagers.

The teenage body hasn't kept up with the demands placed on it. A study by researchers at the University of Kentucky found that the average high school senior sleeps only six and a half hours each night, about three-fourths of what sleep researchers consider necessary for adolescents. Many students find themselves falling asleep in 7:00 a.m. classes no matter how early they try to go to sleep the night before. In one telling example of the impact of early start times, a researcher found that most students earned higher grades in classes that started later in the day for the simple reason that they were more likely to stay awake for the entire lesson.

The lack of sleep affects the teenage brain in similar ways to the adult brain, only more so. Chronic sleep deprivation in adolescents diminishes the brain's ability to learn new information, and can lead to emotional issues like depression and aggression. Researchers now see sleep problems as a cause, and not a side effect, of teenage depression. In one study by researchers at Columbia University, teens who went to bed at 10:00 p.m. or earlier were much less likely to suffer from depression or suicidal thoughts than those who regularly stayed awake well after midnight.

Teenage sleep deprivation appears to be a uniquely American problem. One report found that the average high school in Europe starts at 9:00 a.m., and that far fewer students complain about not getting enough sleep. But back in Minnesota, the first bell at Edina's high school rang at 7:25 a.m.

That was when Edina's school board proposed a solution that was radical in its simplicity. Since students who were awake were more likely to learn something than those who were asleep, the board decided to push the high school's starting time an hour and five minutes later, to 8:30. It was the first time in the nation that a school district changed its schedule to accommodate teenagers' sleeping habits. The response wasn't what the board members expected. Some parents complained that the new schedule would take time away from after-school sports or school clubs. Others said that they needed their children home to babysit their siblings. Yet the most persistent complaint was that pushing the starting time back wouldn't result in better-rested kids, but the opposite. Critics argued that teenagers would simply use the time to stay up even later, compounding the problem and making parents' lives more difficult.

Nevertheless, Edina's teenagers started the 1996–1997 school year on the new, later schedule. The same year, Kyla Wahlstrom became a fixture at district high schools. A former elementary-school principal turned university professor, Wahlstrom conducts research into school policies and how they impact students. She had no prior experience with sleep research, but she knew enough about how schools worked to evaluate the effects of the later starting time. She spent her days interviewing parents, sports coaches, teachers, and students to determine whether the new starting time resulted in a meaningful change or whether it was just the impractical academic theory some called it.

She presented her findings a year later. They were unambiguous. Despite the fears of some parents, teenagers did in fact spend their extra hour sleeping, and reported that they came to school feeling rested and alert. At the same time, the number of on-campus fights fell, fewer students reported feeling depressed to their counselors, and the dropout rate slowed. Coaches pushed back practice times until later in the afternoon, and participation in sports didn't suffer.

The only time that was compromised was the time that teenagers spent being teenagers. "I talked to hundreds of students and the preferred hangout time was always two to four," Wahlstrom told me. "What this has done has severely limited that time. In a sense, they've traded sleeping for hanging out." The effect was quantifiable. The year before the district shifted its starting time, the top 10 percent of students in Edina's high school averaged a combined 1,288 out of 1,600 on their SAT scores. The next year, the top 10 percent averaged 1,500. Researchers couldn't pin the improvement on anything but extra sleep. The head of the College Board, the company the administers that test, called the results "truly flabbergasting."

Following Edina's lead, Minneapolis pushed its high school starting times from 7:15 to 8:40. The two districts were essentially opposites. Edina is an affluent town in which 90 percent of the students are white. In Minneapolis, most students were minorities, and three out of every four teenagers in a classroom came from families whose incomes were low enough to qualify them for subsidized school lunches.

The differences between the urban school district and its suburban counterpart gave Wahlstrom an opportunity to test whether extra sleep did more than just improve the already fortunate lives of wealthy students in a good school district. Just like the previous year in Edina, she became a fixture at Minneapolis's schools, interviewing faculty, parents, and students to collect firsthand data that charted the effects of starting later in the morning. And just like their suburban neighbors, Minneapolis students posted better grades, dropped out less frequently, and attended first-period classes more often following the shift to a later schedule. "The two school districts couldn't have been more different, but there were identical sleep habits between the kids," Wahlstrom told me. "If sleep habits aren't culturally connected, then social, economic, or racial status would have no bearing on them. That patterns were exactly what you would expect if it was biology."

Wahlstrom's research led to a boom in studies on the starting times at schools. Other districts followed suit, and found effects that sometimes went beyond scholastics. In Lexington, Kentucky, for instance, pushing the starting time back led to a 16 percent reduction in the number of teenage car accidents during a year in which teenage accident rates rose 9 percent for the state as a whole. In Rhode Island, pushing starting times back a half hour resulted in a forty-five-minute increase in the average amount of time that the average student spent sleeping. "Our mornings are a whole lot nicer now," the lead researcher of the study, whose daughter was a high school student, said at the time.

Allowing children to get additional sleep may help solve the problem of school bullying as well. A 2011 University of Michigan study tracked nearly 350 elementary school children. About a third of the students regularly bullied their classmates. Researchers found that the children with behavioral issues were twice as likely to have excessive daytime sleepiness or to snore, two symptoms of a persistent sleep disorder. Louise O'Brien, an assistant professor of sleep medicine at the University of Michigan who was part of the research team, argued that "the hypothesis is that impaired sleep does affect areas of the brain. If that's disrupted, then emotional regulation and decision-making capabilities are impaired."

A school superintendent calls Wahlstrom about her research about once a week. When I spoke with her, she was working on a study funded by the Centers for Disease Control and Prevention. In a sign that health professionals are starting to take the issue seriously, the goal of the study will be to determine whether teenage sleep deprivation and early starting times at school amount to a public health issue as serious as smoking and obesity.

And yet, even with the abundance of data from school districts across the nation, a steady number of callers seek out Wahlstrom to question whether changing a school's starting time is really worth it. Many, she says, are school board members or district superintendents who want the same results but don't know how to sell the later starting time to skeptical parents. "A lot of the concern out there has to do with tradition with a capital T," she told me. "We have this Puritan work ethic of early to bed and early to rise. But teenagers can go to bed early and be dog tired, and end up staring at the ceiling until eleven."

10

Breathe Easy

This is the tale of how an Australian man with a vacuum cleaner fixed a mistake in evolution. It begins in the late 1970s. Colin Sullivan is a physician in the Respiratory Unit at Sydney's Royal Prince Alfred Hospital. There, he treats patients who have problems breathing. The most common complaint, by far, is snoring. Sullivan knows better than most doctors in his field that snoring is often a sign of a serious disorder known as sleep apnea. The disorder had been identified only about a decade earlier. Patients with sleep apnea experience a strange nightly sensation that brings the body disturbingly close to death. First, the throat closes randomly throughout the night, choking off the body's air supply. This puts in motion a cascade of increasingly bad side effects. As if on a seesaw, the lack of air causes the oxygen levels in the blood to plummet and the blood pressure to jump. The lips and the skin start to turn blue. Air may not come into the lungs for up to a minute. And for some patients, the heart stops beating for almost ten seconds at a time.

Eventually, the brain gets the urgent message that the body is choking. The brain jolts awake, and the body instinctively gasps for air. Yet as soon as the airway is clear, the brain immediately falls back to sleep. That's when the cycle starts again. It is all so quick that it can happen more than twenty times an hour, all night long, without the sleeper remembering it the next day. Someone lying next to him or her, however, can hear this process at work: when the rhythmic sawing of a snorer's breath pauses and then becomes a hard *ghhack-ghhack-ghhack*, it's most likely the body frantically clearing its airway.

Patients with mild cases of sleep apnea complain of constant exhaustion, a result of never spending more than a few minutes asleep at a time. Severe cases can be life-threatening. A 1992 report by the National Commission on Sleep Disorders estimated that sleep apnea was the cause of thirty-eight thousand fatal heart attacks and strokes in the United States each year.

Sleep apnea was discovered when a group of American physicians noticed that some obese patients complained of overwhelming fatigue and would drift asleep unintentionally. With a literary flourish, they named the condition Pickwickian syndrome after a character in Charles Dickens's first novel, *The Pickwick Papers*, who falls asleep standing up. Doctors incorrectly attributed the sleepiness to a combination of excess weight and abnormally high levels of carbon dioxide in the blood. It was only later that science understood sleep apnea to be a common breathing disorder caused by the position of the tongue and tissues of the throat. It was then given the name *apnea,* from the Greek word for breathless.

Sleep apnea was on the frontier of sleep medicine in the late 1970s. Sullivan had recently returned from a fellowship in Toronto, where he spent three years studying the breathing patterns of dogs while they slept. English bulldogs, pugs, and other breeds with pushed-in faces are the only animals besides humans that experience sleep apnea. The years spent working with dogs gave Sullivan an idea. Once back in Sydney, he devised a mask that fit over a dog's snout. The mask continuously pumped in air from the surrounding room, increasing the air pressure in the throat and preventing it from closing up. Experiments with dogs suggested that the steady flow of air dramatically improved sleep. All Sullivan needed was a human to try it out on.

In June of 1980, he found one. A man walked into the hospital with such a severe case of sleep apnea that Sullivan recommended an immediate tracheotomy. This procedure, which consisted of making a hole in the throat to allow a person to breathe without using the nose or mouth, was one of the few approved treatments for sleep apnea at the time. It required a permanent quarter-size opening in the neck, however, and was quite painful.

The patient refused the tracheotomy. But he was happy to

volunteer as a test patient for Sullivan's air-pressure machine. Sullivan built a test model that afternoon. He grabbed the engine out of a vacuum cleaner and attached it to a handful of plastic tubes. He then took a diving mask and coated the edges with a silicone sealant that prevented air from leaking out of it. Soon, he had a system that allowed him to pump air through the mask at a controlled pressure. Sullivan found an empty room in the hospital and set up equipment to monitor the patient's breathing and brain waves, which would tell him what stage of sleep the man was in. The patient was hooked up to the monitors, put on the mask, and fell asleep almost instantly. He began experiencing sleep apnea within a few minutes. Sullivan then slowly started to increase the pressure in the air flowing through the mask and into the patient's airway. Suddenly, the apnea stopped. The patient began breathing normally. As Sullivan watched in amazement, the patient instantly went into deep REM sleep—a rare phenomenon suggesting that his brain had been starved of restorative sleep. Sullivan then slowly decreased the pressure of the air flowing through the mask. The apnea returned. Sullivan rapidly went through several cycles of increasing and decreasing the pressure. He found that with the machine's controls alone, he could effectively turn the patient's sleep apnea on and off.

The machine worked. The next question was whether its benefits would last all night. Sullivan left the settings on the machine at a level where the patient was free of sleep apnea. Then he waited. For about seven straight hours, the patient was in abnormally intense, deep sleep. When he woke up the next day, he told Sullivan that he felt awake and alert for the first time in years.

Sullivan began searching for other patients willing to serve as guinea pigs. He found five whose long histories of excessive daytime sleepiness and loud snoring seriously affected their lives. Two told Sullivan that they had lost their jobs because of their sleepiness. One subject, a thirteen-year-old boy, had been classified as mentally retarded after he was unable to stay awake at school. Sullivan observed each patient for three nights in a sleep lab. On the third night, he had each test the mask. Just as with the patient in the hospital, the positive airway pressure prevented their throats from closing while they slept. Patients told him that the improvement in their sleep was life-changing.

But Sullivan's mask wasn't embraced in the medical field as quickly. Many doctors were not convinced that sleep apnea was a serious condition, and even fewer thought that a person would be willing to sleep wearing a mask night after night. One told Sullivan that his machine was nothing more than a moneymaking fad. Sullivan continued to refine the mask, experimenting with ways to give it a tighter seal on the face without making it more uncomfortable. With the help of an engineer from the University of Sydney, he began crafting masks that featured various shapes of the nose. He experimented with motors to cut down on the noise, pulling one from a paint compressor and another from a different-model vacuum cleaner. Patients began coming to him from all over Australia. One man, a truck driver, was like many of the patients Sullivan treated: he fell asleep sitting up while talking with Sullivan, and woke up only when he began thrashing his legs in his sleep. When asked about it, he admitted that he had been doing the same thing for more than twenty years. By 1985, Sullivan had more than a hundred patients using a continuous positive airway pressure device on a long-term basis.

The next year, Sullivan met a former university professor and fellow Australian named Peter Farrell. At the time, Farrell had recently given up a job studying kidney disorders at the University of Washington to become a business consultant for Baxter International, which by 2011 was a $30 billion health care company. He was on the lookout for new medical devices. Sullivan said that he had one. Together, they watched clips that Sullivan had filmed of his patients before and after turning his machine on. In one clip, a man lay on his back snoring loudly. Suddenly his breathing stopped, a sign that his upper airway had closed. Monitors on the screen showed the patient's heart rate and blood pressure going haywire. Forty seconds later, the airway cleared and the man took another breath. His heart rate and blood pressure instantly spiked. Sullivan finally turned to Farrell and asked, "Do you think that's good for him?"

Farrell asked to talk with patients who were using the machine. They told him that they slept with it every night despite the drawbacks, chief among them being the noise. The machine still ran off of a vacuum-cleaner engine, creating a whirl so loud that one tester told Farrell that he had cut a hole in his bedroom wall so he could leave the base of the machine in the next room. Another patient who met with Farrell had bruises in an oval pattern over his face from the suction caused by the mask, but still wore it every night despite them.

Farrell did a rough calculation in his head: At Baxter, kidney disease accounted for $2 billion a year in revenues, even though it affected just two people out of a thousand. If sleep apnea affected just one out of a hundred snorers, it would be a $100 billion business. Farrell and Sullivan raised $600,000 to commercialize Sullivan's machine. The new company, called ResMed, introduced its first continuous positive airway pressure device, known as a CPAP, into the market in 1989. Within five years, the company was making $300 million a year in revenue.

On a perfect summer day in San Diego, I pulled up to an eight-story glass building next to a private airstrip. Freshly planted trees tied to plastic stakes lined the walkway to the main entrance. A few years ago, this land was one of the last undeveloped parcels within the city limits. Now, it was the headquarters of ResMed's global business. In the years since the company was founded in Australia, Sullivan's CPAP machine had become the standard treatment for sleep apnea. Four of every ten patients with sleep apnea in the United States were using a ResMed device when I visited the company. Orders for six hundred thousand new machines a month were coming in. The company had the top-selling CPAP machines in Europe, and was growing its business lines in China and India.

The growth rate had caught the attention of Wall Street. Just a few weeks before I visited ResMed's headquarters, Jim Cramer, CNBC's unhinged stock picker, had singled out the company's stock as one of his weekly choices. "This company is the only pure play on sleep out there!" he screamed. The stock gained several percentage points that day. The day before I arrived, ResMed announced that its sales had grown to $1.1 billion a year. It was in the midst of a fifteen-year streak in which revenues and profits had risen every quarter. As I walked into the company's global headquarters, employees were just finishing up a party to celebrate. Balloons and the smell of barbecue filled the lobby.

A collection of ResMed's breathing masks lined the wall. They came in a variety of shapes and sizes, from one that resembled the face mask of a fighter-jet pilot to a small mask designed to fit a four-year-old. One sleek, pink model built for a woman was so thin that it looked like a garden hose. The assortment of models hinted at the fact that snoring, and sleep apnea, aren't limited to obese patients as originally thought. A study in 1994 found that about 10 percent of women, and 25 percent of men, have difficulties breathing in their sleep. These numbers climb as a person gets older, so that as many as one out of three elderly men have at least a mild case of sleep apnea. All told, about twenty million Americans have the disorder.

Its cause could simply be the trade-off that the human body makes for having the ability to speak in a complex language. A short tour of fossils illustrates this point. If you were to look at a Neanderthal's mouth, you might think that its descendants would have been the ones to survive over the long run, considering their jawbones were larger and stronger than our own. Plus, with extra room in their mouths, Neanderthals never experienced the pain of impacted wisdom teeth. *Homo sapiens* differed from Neanderthals by developing a flatter face, a smaller jawbone, and a tongue that descends deeper into the throat than in any other mammal. With this new hardware, humans were able to move beyond making simple grunts. Those first, complicated sounds uttered by *Homo sapiens* soon developed into language. Jared Diamond, a professor at UCLA, called the positioning of the tongue our greatest evolutionary advantage. "It's easy to appreciate how a tiny change in anatomy resulting in capacity for speech would produce a huge change in behavior," he noted. "With language, it takes only a few seconds to communicate the message, 'Turn sharp right at the fourth tree and drive the male antelope toward the reddish boulder, where I'll hide to spear it.' Without language, two proto humans could not brainstorm together about how to devise a better tool or about what a cave painting might mean. Without language, even one proto human would have had difficulty thinking out for himself or herself how to devise a better tool."

But the positioning of the tongue in the *Homo sapiens* mouth complicates the acts of eating, drinking, and breathing. Food could literally go down the wrong pipe, a biological problem unique to modern humans. Darwin noted "the strange fact that every particle of food and drink we swallow has to pass over the orifice of the trachea with some risk of falling into the lungs." The longer tissues of the soft palate at the back of the throat made it possible for the airway to become blocked after a routine exhalation, which could start the cycle of sleep apnea. In the mid-1990s, researchers in Japan found that slight changes in the size and position of the pharynx at the back of the throat drastically increased the likelihood that someone would develop a breathing disorder during sleep. The shape of a person's neck and jaw can also be a factor. A large neck, tongue, or tonsils, or a narrow airway often signal that a person will develop sleep apnea because of the increased chance that breathing will become blocked during the night.

And yet the physicians who first recognized sleep apnea were half right when they assumed that the disorder was a side effect of obesity. Sleep apnea is a flaw that is part of the blueprint of the human body, and excess fat often teases it out. The chances of developing sleep apnea go up with weight because the tissues in the throat become enlarged, making it more likely that they will obstruct the airway during sleep. For some patients, losing weight alone can solve the problem. Other changes in behavior—like drinking less alcohol, cutting back on smoking, sleeping on one's side instead of on the back, or doing exercises or playing musical instruments that build up the muscles in the throat—can also help.

Breathing masks like ResMed's are the most common medical treatment for sleep apnea, but they aren't for everyone. Some patients never get used to the awkward sensation of sleeping with a mask on their face, or never become comfortable with breathing in the cold air that is continuously pumped into their mouth throughout the night. In the long term, patients with mild sleep apnea wear the masks between 40 and 80 percent of the time, according to various studies. There is also a social stigma that complicates treatment. Some patients with sleep apnea decide not to use a CPAP machine because they are worried that it will make them less attractive to the person they are sharing a bed with. In an online support group for patients with sleep apnea, a man wrote that he was "feeling like I am going to be Darth Vader if I have to wear one." A woman wrote that her husband "fought it, cried, said he is defective, said he would prefer to put a gun to his head then wear one of those things." Another wrote that "I've yelled that I feel like a freak to my husband way too many times this fall."

Dental devices are typically the next choice. These aren't as effective as CPAP machines for severe sleep apnea, but they may be easier for some patients to use, especially those who have to travel frequently. One of the most popular looks like a sports mouthguard. It forces the lower jaw forward and slightly down to keep the airway open. Another device holds the tongue in place to prevent it from getting in the way. Surgery is the last option. One procedure, called a uvulopalatopharyngoplasty, consists of removing excess soft tissue from the back of the throat. Its long-term success rate is only about 50 percent, and it can lead to side effects such as difficulties swallowing, an impaired sense of smell, and infection. It is also extremely painful. Few medications have been shown to help sleep apnea, and may in fact make the problem worse. Sleeping pills and tranquilizers, for instance, can make the soft tissues in the throat sag and obstruct the airway more than they would otherwise.

I made my way to Peter Farrell's office in the ResMed building. He sat behind his desk, oval glasses perched on his nose, and stared at me with the intensity of a boxer. He had made a fortune from his work with Sullivan. Still, he thought that sleep apnea remained poorly understood and underrecognized in the United States. "We are still in the early phases of a monster area," he told me, in a thick Australian accent. "This is arguably the biggest health problem in the country and we think that three in ten adults have it. There isn't anything remotely close to that. We have so much runway ahead of us that it's like we haven't even started."

Much of ResMed's growth had come since 2000. That year, four separate studies found conclusive evidence that sleep apnea was associated with increased rates of hypertension. Left untreated, patients with sleep apnea are at a greater risk of developing kidney disease or vision problems, or having a heart attack or stroke. Those studies helped convince government insurance programs such as Medicare, Medicaid, and the British National Health Service to pay for a portion of the cost of each device, which can be several-thousand dollars if a patient were to buy it out of pocket. Sleep labs across the country now conduct overnight tests in which patients who are suspected to have sleep apnea are hooked up to equipment that monitors their hearts, breathing patterns, and brain activity, as well as the number of times they wake up throughout the night and how often they move their limbs.

As scientists began to understand sleep apnea in more depth, they started to see it as the foundation for serious illnesses affecting the mind. In one study, researchers at UCLA conducted brain scans of patients with long histories of sleep apnea and compared them with the scans of control subjects who had normal sleep patterns. The investigations focused their inquiry on the mammillary bodies, two structures on the underside of the brain so named because they resemble small breasts. Mammillary bodies are thought to be an important part of the memory and have long been associated with cases of amnesia. This memory center of the brain was 20 percent smaller in patients with sleep apnea. Had a doctor looked at a patient's brain scan alone, it would have suggested severe cognitive impairment: a similar shrinkage in the size of the mammillary bodies is found in patients with Alzheimer's disease or those who experienced memory loss as a result of alcoholism. It was the first indication that sleep apnea leaves a permanent scar beyond the daily difficulties of focus and attention that come with sleepiness. "The reduced size of the mammillary bodies suggests that they suffered a harmful event resulting in sizable cell loss," noted Ronald Harper, professor of neurobiology at the David Geffen School of Medicine at UCLA and the lead investigator for the study. "The fact that patients' memory problems continue despite treatment for their sleep disorder implies a long-lasting brain injury."

A study published in the *Journal of the American Medical Association* supported this conclusion. Kristine Yaffe, a professor of psychiatry at the University of California, San Francisco, led a study that recruited nearly three hundred elderly women who were mentally and physically fit. The average age of the subjects in the study was eighty-two. Each woman spent a night in a sleep lab, and Yaffe found that about one in every three met the standard for sleep apnea. Yaffe reexamined each woman five years later. The effects of age on the mind seemed to depend on the quality of sleep. Nearly half of the women with sleep apnea showed signs of mild cognitive impairment or dementia, compared with only a third of the women who slept normally. After controlling for factors such as age, race, and the use of medicines, Yaffe found that the women with sleep apnea were 85 percent more likely to show the first signs of memory loss. The frequent interruptions in sleep, and the reduced oxygen in the brain, may reduce the brain's ability to form and protect long-term memories.

Sleep apnea's effects on the brain can also have devastating consequences on the highway. By any measure, commercial truckers have a difficult job. Confined to one position and forced to maintain attention for a long time while racing to meet constant deadlines, many truckers wear the signs of stress on their bodies. Stefanos N. Kales, an assistant professor at Harvard Medical School and the Harvard School of Public Health, began tracking the outcome of a truck driver's lifestyle, including poor nutrition, little exercise, and less sleep. Obesity was widespread and contributed to a much higher rate of sleep-disorder breathing than what is found in the general population. Previous studies suggested that about one out of every three big-rig truckers had moderate to severe sleep apnea, a rate indicating that thousands of drivers were straining to stay awake on the road. By Kales's estimates, a driver with sleep apnea was seven times more likely to get in an accident. More disturbingly, Kales found that one out of every five accidents involving a commercial truck was caused by its driver falling asleep at the wheel.

Drivers are rarely willing to admit that they have sleep

apnea, much less seek treatment, because doing so could increase their chances of losing their commercial licenses and livelihoods. Kales led a study in which his team observed nearly five hundred truckers from fifty different companies over a period of fifteen months. Screening questionnaires flagged about one in every six drivers as showing signs of probable sleep apnea. Of these, only twenty drivers agreed to spend a night in a sleep lab. All were shown to have the disorder. And yet only one driver out of the group began treatment, using a CPAP device regularly. "Screenings of truck drivers will be ineffective unless they are federally mandated or required by employers," Kales and his team members noted.

The Government Accountability Office, the nonpartisan research arm of the federal government, found that the process of certifying commercial drivers routinely overlooks serious health concerns that could impact a driver's ability. Its report identified more than half a million truckers nationwide who had a commercial driver's license at the same time that they were eligible to receive full disability benefits from the federal government. Sleep apnea was common and untreated, and it had fatal consequences. In July of 2000, the driver of a big rig rammed a Tennessee Highway Patrol car that was protecting a highway work zone. The patrol car exploded upon impact, taking the life of a state trooper. The driver of the big rig had been diagnosed with sleep apnea, but he wasn't treating his disorder. Nor was it his first accident. Three years earlier, he had struck a patrol car in Utah. Five years later, another trucker with a severe form of sleep apnea collided with a sports utility vehicle in Kansas. Its passengers, a mother and her ten-month-old baby, were killed. Like in Tennessee, the driver of the big rig had been diagnosed with sleep apnea. But in order to get his medical clearance, he went to a doctor who had never treated him before and did not disclose his illness. The driver was later found guilty of two counts of vehicular manslaughter. One federal proposal would require sleep apnea screenings for drivers whose body mass index exceeds 30, the baseline number for obesity. Truckers have been vocal in their opposition. "There is no direct relationship between a person's body weight and his ability to drive an 18-wheel truck," said a spokesperson for an organization that represented about 160,000 commercial truckers. "Show me where that's a better predictor than a person's driving record."

Sleep apnea and weight are not problems limited to the United States, a fact that hasn't been lost on companies like ResMed. Kieran Gallahue, CEO of the company at the time of my visit in 2010, came to ResMed after holding a series of positions at Procter & Gamble and General Electric. In his new role, he retained the no-nonsense air of a rising executive at a blue-chip company. He brought me into his office. With the flick of a switch behind his desk, a white board rose up from a bookshelf next to him. He began diagramming the company's long-term strategy, sprinkling in phrases like "We are going through our organizational adolescence" that would have made his professors at the Harvard Business School proud. Gallahue argued that ResMed's breathing machines have the ability to prevent serious illness, and in the process drive down the costs of health care. “We’re a solution to the cost containment problem,” he said.

At the same time, however, the company was counting on worldwide obesity rates to continue to rise. The spread of Western fast-food companies like McDonald’s, Kentucky Fried Chicken, and Pizza Hut to emerging countries such as China and India may be the greatest growth engine for ResMed. Simply put, more fat in the bodies of the world’s population equals a larger number of sleep apnea cases, creating a larger customer base for ResMed’s products. “Genetically you’re still engineered for a low-calorie, low-fat diet,” Gallahue told me. “That’s what your body has been optimized for over centuries. Boom, you introduce burgers, and your body is not going to handle it. One of the outcomes is going to be a skyrocketing in the prevalence of sleep disordered breathing.”

Everywhere I turned in ResMed’s headquarters were signs of a company trying to stay ahead of itself. Gallahue’s office overlooked what appeared to be a lush private park for ResMed’s employees. A group of workers was eating lunch at one of many clusters of benches spread over a space larger than a football field, lined with a jogging trail and decorative footlights. Metal sculptures glimmered in the sun. I asked Gallahue if setting up the outdoor area was meant to help the employees’ quality of life and morale. He paused for a second as he considered my question and then laughed. “Actually, that’s land that we plan to build a second building on,” he told me. “The park is only temporary.”

11

Counting Sheep

A 1945 U.S. Navy training film begins by showing a room full of sailors watching a cartoon. The men literally hoot and holler as Donald Duck tries, and repeatedly fails, to fall asleep. First, Donald misjudges the location of his pillow and slams his head down onto a metal bed frame. Next, his alarm clock begins ticking so loudly that it shakes the nightstand. Donald becomes enraged and smashes the clock into the wall. Finally, just when it seems like everything is calm and he lays his head back on the pillow, his Murphy bed snaps shut with him inside it. A sailor named Lucky laughs so hard at Donald's misfortune that he has to dab his eyes with a handkerchief.

But Bunce, a curly-haired sailor sitting next to Lucky, watches the film with sullen, unblinking eyes. While the men around him laugh and joke with each other, he broods, finding no humor in Donald's losing battle to get to sleep. The cartoon ends, and Lucky and Bunce turn to leave. Lucky asks Bunce why he wasn't enjoying the film along with the rest of the group. "What's so funny?" Bunce snaps back at him. "You're one of those slap-happy guys that sleeps like a babe."

We soon learn that Bunce suffers from insomnia, and he doesn't find it a laughing matter. Lucky decides that he can cure his buddy's sleep problem before the night is over. As the men brush their teeth, he tells Bunce to stop worrying about a girl who hasn't written him for three weeks. Standing in the shower, he reminds Bunce that everyone gets sleepless sometimes. No one in their unit slept while they were preparing for the attack on Saipan, for instance. And as they head to bed, Lucky tells Bunce that he should tell him if anything is on his mind.

None of it works. As Lucky sleeps with a smile on his face, Bunce lies in bed, staring at his wristwatch as the seconds tick by. The camera zooms in on his pouting face, and we hear a voice-over of Bunce's increasingly frantic internal monologue. "Go to sleep. Go to sleep! Why can't I get some sleep? Night after night. I can't take it. I'd rather be dead. Probably will die. Nobody, nobody can last without sleep night after night. Night after night after night."

That's when the voice of a friendly off-screen doctor chimes in. "Oh no," he says, with a slight down-home twang and laugh.

"We know how you feel. But in all of medical history, nobody has ever died from lack of sleep."

The reminder that insomnia isn't fatal is a small comfort to someone experiencing it. Every night, about two of every five adults in the United States have problems falling and staying asleep that aren't related to a persistent sleep disorder. As they lay in bed, many are caught in the classic paradox of insomnia: wanting sleep so badly that they can't get it. "The condition of sleep is profoundly contradictory," noted Emily Martin, a professor at New York University who has studied insomnia. "It is a precious good . . . but it is a good like none other, because to obtain it one must seemingly give up the imperative to have it." The psychologist Viktor Frankl noted in 1965 that "sleep [is like] a dove which has landed near one's hand and stays there as long as one does not pay any attention to it; if one attempts to grab it, it quickly flies away."

For doctors, insomnia presents a bit of the chicken or the egg problem. Is the sleeplessness a result of another condition, such as depression, or is the insomnia the root of the other problem? One report by the National Institutes of Mental Health found that depression rates were forty times higher for patients with insomnia than those without sleep problems. Mental health experts increasingly view depression or anxiety as an effect, rather than a cause, of insomnia. Taking care of insomnia may therefore calm other aspects of a patient's life.

And yet insomnia is a unique and difficult condition to treat because it is self-inflicted. The cause is often the brain's refusal to give up its unequaled ability to think about itself, a meta-phenomenon that Harvard professor Daniel M. Wegner has called "the ironic process of mental control." To illustrate this concept, imagine someone telling you that you will be judged on how quickly you can relax. Your initial reaction most likely is to tighten up. After he posed that challenge to research subjects, Wegner found that the average person becomes anxious as his or her mind constantly monitors its progress toward its goal, caught up in the second-by-second process of self-assessment. In the same way, sleep becomes more elusive as a person's sleep needs become more urgent. This problem compounds itself each night, leading to a state of chronic insomnia.

Wegner demonstrated how the mind's ironic sense of control played out in the real world of sleep. He sent 110 undergraduates home with a Walkman, a cassette, and instructions to listen to the tape as soon as they got into bed and turned the lights off. Each student was a normal sleeper, without any history of insomnia or another chronic sleep disorder. As they lay in bed, half of the test subjects heard this message: "Good evening . . . As you listen to the music that follows, you should try to fall asleep as quickly as possible. Your task is to put yourself to sleep in record time. Please concentrate on going to sleep quickly." The other group, meanwhile, heard essentially the opposite. "Your task is to fall asleep whenever you would like."

The design of the experiment included a second tier of anxiety. Ninety minutes of music followed the instructions. Half of the subjects who had been told to fall asleep as soon as possible heard the blaring of a loud marching band. Wegner chose this music to give subjects an additional mental hurdle to pass. Not only had they been given a deadline, but also they now had to question whether it would ever be possible to meet it in the first place. The other half of the subjects in the study heard what the experiment described as "new age music . . . containing restful outdoor sounds such as birds, crickets, and a stream bubbling in the background." An equal number of subjects who had been told to fall asleep whenever they wished heard either the marching band or the crickets as well.

As predicted, subjects who had been told to fall asleep quickly took longer to do so. Their minds were so focused on falling asleep in record time that they found themselves consciously checking on their progress, unable to let their thoughts drift off and guide them to dreamland. And not surprisingly, those who were trying to fall asleep urgently while listening to the taxing music of the marching band fared the worst by far.

But here was where the study defied expectations. The misfortune of the fall-asleep-as-quick-as-you-can group wasn't just limited to time spent listening to the marching band music. Throughout the night, these subjects woke up more often, and had a harder time getting back to sleep, than any other group, even after their headphones went silent. The next day, they reported feeling less rested than their peers. The stress of trying to fall asleep while the music was playing had lingered well into the early morning. Just like Bunce in the military training film, they wanted to sleep so badly in those first minutes in bed that they couldn't calm their minds down throughout the night. Wegner had set in motion the cycle of insomnia.

Treating insomnia isn't easy. Part of the reason is the fact that science, as a whole, has a fuzzy definition of what constitutes the disorder. One night of bad sleep because of a blaring car alarm or an upcoming stressful day at work doesn't classify as insomnia. Instead, it is generally thought of as a string of otherwise peaceful nights during which a patient can't fall asleep when he or she wants to. The National Institutes of Health identifies the condition as "difficulty getting or staying asleep, or having non-refreshing sleep for at least one month." The classic form of short-term insomnia has no known cause yet is widespread. About one in ten people in the United States suffer from it during their lifetime.

There is no medical test that proves whether someone is suffering from a temporary bout of sleepless nights or a more serious disorder. Some patients go to sleep labs and undergo tests to rule out conditions such as sleep apnea, but knowing what they don't have offers little help in treating what they do. Instead, doctors rely on self-reports from patients, which can be maddeningly vague, a result of the difficulty that we have with accurately noting how many hours we truly spent sleeping on any given night. Patients who have spent a night in a sleep lab, for instance, often complain that it took them more than an hour to fall asleep when a chart of their brain waves shows they were asleep within ten minutes. Problems of self-reporting aren't limited to judging how long it took to get to sleep. Some patients wake up in labs claiming that they didn't sleep at all during the night, despite hours of video and brain wave evidence to the contrary.

It is a part of the paradox that sleep presents to a conscious mind. We can't easily judge the time that we are asleep because that time feels like an absence, a break from the demands of thought and awareness. The times that we do remember are those that we wish we couldn't: staring at the clock in the middle of the night, turning the pillow over desperately hoping that the other side is cooler, kicking the covers off or pulling them up close. Those experiences, even if they last only three minutes, often become exaggerated in our minds and overshadow the hours that we spent sleeping peacefully, simply because we remember them.

When insomnia starts to interfere with the routines of normal life, many people decide to turn to pharmaceuticals. Medicines that help someone fall asleep, stay asleep, or be comfortable in between accounted for $30 billion in annual sales by 2010 in the United States alone, which is a little more than what people around the world spend each year going to the movies. Sleeping pills are responsible for the majority of those profits. It is a remarkable turnaround, considering that it wasn't that long ago that public distrust of sleeping pills led *Coronet* magazine, a spin-off *Esquire* published until the mid-1960s, to call them "the doorway to doom."

In 1903, a physician named Joseph von Mering and a chemist named Emil Fischer developed the first modern medication that promised a safe way to induce sleep. Von Mering had made a name for himself fifteen years earlier when he discovered that the pancreas was responsible for the production of insulin, an important leap forward in the treatment of diabetes. To determine exactly what the pancreas did, von Mering decided to open up his dog, cut out the organ, and see what happened. The dog survived the surgery and undertook a revenge that was all too short-lived. Though house-trained, the dog began to urinate in von Mering's lab. This happened so often that von Mering decided to have the urine tested. He found that it had high levels of sugar, one of the telltale signs of diabetes.

In an early attempt at branding, von Mering and Fischer called the sleeping pill they developed Veronal, hoping that the name would play off the image of the city of Verona as a place of peace and quiet. The new drug belonged to a class of medications known as barbiturates, which, when taken in low doses, often make a patient feel intoxicated. While the pills did allow some patients to at least temporarily reach what Veronal advertisements described as "natural sleep," they came with some serious side effects. Chief among them was that the body easily developed a tolerance for the drug, making a patient require progressively larger doses for it to work.

This wouldn't have been so bad had the recommended dosage of the pills not been so close to a fatal one, especially when mixed with a little alcohol. For the next sixty years, sleeping pills were blamed for countless accidental overdoses when patients took an extra pill or two in a half-asleep daze. Brian Epstein, the manager of the Beatles, died in his London home after taking a lethal dose of barbiturate pills. The death was officially ruled an accident. Their availability and potency also made barbiturate sleeping pills a factor in a number of well-publicized suicides. Actor Grant Withers, who appeared in a string of John Wayne films, turned to sleeping pills when he took his own life in 1959. A bottle of barbiturate sleeping pills was found next to the body of Marilyn Monroe three years later. The popularity of the drugs plummeted after the deaths in Hollywood, as both doctors and patients were spooked to realize that the pills in their bathrooms were capable of killing so easily.

A family of sedatives called benzodiazepines became popular in the 1970s because they were thought to be safer than their predecessors. This type of drug, which includes variants such as Valium (diazepam) and Rohypnol (flunitrazepam), work by binding to the receptors in the brain that arouse a person out of sleep, essentially making it harder for him or her to wake up. While these pills were an improvement over barbiturates because they drastically lowered the chance of an overdose, the high they gave some patients made them more likely to be abused. That wasn't all. In the late 1980s, patients started to show signs of memory loss after taking a benzodiazepine known as Halcion (triazolam).

The worst of the conditions linked to Halcion was something called traveler's amnesia, which often took place during international travel. Typical patients who experienced traveler's amnesia would take a dose of Halcion while on a red-eye flight, to ease the adjustment to the time difference. When they woke up at their destination, however, their memory would be blank. Patients lost track of who they were, where they had landed, and why they were there. Others would wake up in their hotel rooms and realize that they had no memory of their plane landing, of walking through customs, or of riding in the cab that picked them up from the airport. The United Kingdom banned the drug in the early 1990s, while several other countries severely restricted its availability (it is still legal in the United States).

The sleeping pill market changed in 1993 when a French company now known as Sanofi introduced a new drug called Ambien, also known by its generic name *zolpidem*. Ambien worked in essentially the same way as the benzodiazepines, though with far fewer side effects. It appeared safe enough, in fact, that many doctors broke their long-standing refusal to prescribe a medication for run-of-the-mill insomnia. Ambien quickly dominated the sleeping pill market and rang up more than a billion dollars in sales a year. At one time, Ambien accounted for eight out of every ten sleeping aids prescribed in the United States, a near monopoly enjoyed by few other drugs in history.

It wasn't until 2005 that its first real competitor emerged. That was when a small biotech company in Marlborough, Massachusetts, called Sepracor introduced Lunesta, also known as eszopiclone. Though in the same class of drugs as Ambien, Lunesta had two advantages over it. One was the fact that the FDA approved it for long-term use, which meant that patients weren't advised to forgo taking the drug every couple of days like they were with Ambien. The second was a branding campaign that featured a little green moth that floated onto the faces of happy, smiling actors pretending to be asleep in the company's commercials. "The word 'nest' is hidden in Lunesta so people think of their nests when they sleep," one brand consultant said while praising its launch. Sepracor made sure that everyone saw its moth by spending $230 million on advertising the year Lunesta was introduced, making it the most-promoted drug of the year.

All told, sleeping pills accounted for more than $1 billion in advertising between 2005 and 2006. The sheer number of commercials may have caused as much insomnia as the drugs treated. Just like in Wegner's instructions to his test subjects to fall asleep as quickly as they can, constant reminders and advertisements about obtaining good sleep would be enough to push anyone into the cycle of insomnia, beset by worries over whether his or her sleep measured up to what the commercials offered. In one year, the total number of sleeping pill prescriptions written in the United States jumped from 28 million to 43 million. Every week, 120,000 new patients asked their doctors for a sleeping pill, a growth rate that rivaled the spread of Facebook. After it brought in almost $100 million in sales in its first quarter on the market, Wall Street analysts declared that Lunesta could do for the insomnia market what Prozac did for depression. *Brandweek* awarded Sepracor its Marketer of the Year Award and proclaimed that, thanks to the company, "insomnia is sexy again"—though, to be fair, it was never technically sexy in the first place. By 2010, about one in every four adults in the United States had a prescription sleeping pill in their medicine cabinets.

But here's the twist. A number of studies have shown that drugs like Ambien and Lunesta offer no significant improvement in the quality of sleep that a person gets. They give only a tiny bit more in the quantity department, too. In one study financed by the National Institutes of Health, patients taking popular prescription sleeping pills fell asleep just twelve minutes faster than those given a sugar pill, and slept for a grand total of only eleven minutes longer throughout the night.

If popular sleeping pills don't offer a major boost in sleep time or quality, then why do so many people take them? Part of the answer is the well-known placebo effect. Taking any pill, even one filled with sugar, can give some measure of comfort. But sleeping pills do something more than that. Drugs like Ambien have the curious effect of causing what is known as anterograde amnesia. In other words, ingesting the drug essentially makes it temporarily harder for the brain to form new short-term memories. This explains why those who take a pill may toss and turn in the middle of the night but say the next day that they slept soundly. Their brains simply weren't recording all those fleeting minutes of wakefulness, allowing them to face each morning with a clean slate, unaware of anything that happened over the last six or seven hours. Some sleep doctors argue that this isn't such a bad thing. "If you forget how long you lay in bed tossing and turning, in some ways that's just as good as sleeping," one physician who worked with pharmaceutical companies told the *New York Times,* voicing what is a widely held opinion among the sleep doctors and physicians that I spoke with.

Serious problems can arise when people taking a drug like Ambien don't actually stay in bed. Some have complained of waking up the next day and finding things like candy wrappers in their beds, lit stoves in their kitchens, and bite marks on the pizzas in their freezers. Others have discovered broken wrists that came from falling while sleepwalking, or picked up their cell phones and seen a list of calls that they have no memory of making. Ambien was part of a kinky footnote in the Tiger Woods saga. One of his mistresses said that the pair would take the drug before sex because it would lower their inhibitions. Visitors to Sleepnet, an Internet forum, have noted their own troubles with sleeping pills. "Many people have tried to convince me that Ambien is a good drug. Maybe it works for some people, but I have to tell you it has been one big nightmare for me and my family," one person wrote. "I have done the most dangerous and humiliating things after taking the drug. To provide some examples, I have called people, who I never would have called and said things that I would never have said, leading to very uncomfortable relationships and explanations of why I called. I have had sexual encounters that I barely remember. I have left my apartment in pajama-like attire to go shopping in the middle of the night. Once I wrote all over the walls of my apartment with nail polish. That was a nightmare." Not long after a member of the Kennedy family blamed a car accident on the effects of Ambien, the Food and Drug Administration issued new rules that require pharmacists to explain the risk that taking certain sleeping pills could lead to things like sleep eating, sleepwalking, or sleep driving.

Those warnings have done little to tamper with the popularity of sleeping pills, especially since the most popular one is cheaper than ever. Ambien went off patent a few months before the FDA issued its new requirements. While the total number of dollars spent on sleeping pills fell by more than $1 billion a year because of the availability of cheaper generic versions, the number of patients filling a prescription for them remained steady. Many people who take sleeping pills find that their sleep quality reverts to its previous, poor state the night they decide to go without medication, a vicious cycle that increases their dependency on a drug approved only for short-term use. Facing a night of sleep without backup produces the same form of stress that originally caused the insomnia cycle to begin.

Yet there is a way to treat insomnia without setting patients up for a letdown as soon as the prescription runs out. Charles Morin is a professor of psychology at Université Laval in Quebec. For more than ten years, he has studied whether modifying behavior can be as effective at treating insomnia as taking medication. His research focuses on a type of counseling called cognitive behavioral therapy, a treatment that psychologists often use when working with patients suffering from depression, anxiety disorders, or phobias. The therapy has two parts. Patients are taught to identify and challenge worrisome thoughts when they come up. At the same time, they are asked to record all of their daily actions so that they can visualize the outcome of their choices.

When used as a treatment for insomnia, this form of therapy often focuses on helping patients let go of their fear that getting inadequate sleep will make them useless the next day. It works to counter another irony of insomnia: Morin found that people who can't sleep often expect more out of it than people who can.

Patients with insomnia tend to think that one night of poor sleep leads to immediate health problems or has an outsized impact on their mood the next day, a mental pressure cooker that leaves them fretting that every second they are awake in the middle of the night is another grain of salt in the wound. In the inverted logic of the condition, sleep is extremely important to someone with insomnia. Therefore, the person with insomnia can't get sleep.

In a study in 1999, Morin recruited seventy-eight test subjects who were over the age of fifty-five and had dealt with chronic insomnia for at least fifteen years. He separated his subjects into four groups. One group was given a sleeping pill called Restoril (temazepan), a benzodiazepine sedative often prescribed for short-term insomnia. Another group was treated with cognitive behavioral therapy techniques that focused on improving their expectations and habits when it came to sleep. The members of this group were prompted to keep a sleep diary and meet with a counselor to talk about their patterns, as well as carry out other actions. The third group was given a placebo, and the fourth was treated with a combination of Restoril and the therapy techniques.

The experiment lasted for eight weeks. Morin then interviewed all of the subjects about their new sleeping habits and the quality of their sleep each night. Patients who had taken the sleeping pill reported the most dramatic improvements in the first days of the study, sleeping through the night without spending any of the lonely hours awake they had come to expect. Subjects who were treated with the cognitive behavioral therapy began to report similar results in sleep quality a few days later. Over the short term, sleeping pills had a slight edge in sanding down the rough edges of insomnia.

But then Morin did something extraordinary in the field of insomnia studies: after two years, he contacted all of his subjects and asked them about their sleeping habits again. It was a novel approach to investigating a disorder that often appears solved as soon as a patient sleeps normally for a few nights. Morin wanted to determine whether sleeping pills or therapy would do a better job of reshaping the underlying causes of persistent insomnia. Subjects who had taken the sleeping pills during the study told him that their insomnia returned as soon as the drugs ran out. But most of those who went through the behavioral therapy maintained the improvements they had reported in the initial study. Lowering patients' expectations of sleep and helping them recognize what contributed to their insomnia combined to be more powerful over the long run than medication. "In the short run, medication is helpful," Morin told the *New York Times*. "But in the long run, people need to change their actual sleep habits — that's what [therapy] helps them do."

Therapy is also helpful at breaking a person's reliance, either real or imagined, on sleeping pills. In a 2004 study, Morin found that nine of every ten subjects who combined a gradual reduction in their medication with cognitive behavioral therapy were drug-free after seven weeks. Only half of those who tried to stop using the pills by reducing dosage alone were as successful. Further tests revealed that subjects who relied on therapy experienced better sleep quality as well, with longer amounts of time in deep sleep and REM sleep. A separate study the same year found that one of every two subjects who began a cognitive behavioral treatment plan no longer felt the need to take sleeping pills. The results from these and other cognitive behavior therapy studies have been compelling enough that organizations ranging from the National Institutes of Health to *Consumer Reports* recommend therapy as the first step in treating insomnia.

Advice that comes remarkably close to cognitive behavioral therapy is what ultimately helps Bunce, the sleep-starved sailor. The off-screen doctor tells Bunce that he needs to channel the energy he spends worrying about being awake into improving his ability to relax. "Get this first," the doctor says. "Relaxing is a skill, like hitting a target. It takes practice, concentration, and more practice." Bunce is then taught the basic skills of progressive relaxation therapy. He is told to first release the tension in his feet. Then, to stretch out his legs and sink into the bed. He continues to relax his body, unfurrowing his brow and unclenching his jaw, all as part of a strategy to convince his mind to let up from its intense focus on sleep.

Yet some people with insomnia may never respond to therapy like this, simply because their sleeplessness isn't a reflection of the mind putting pressure on itself. Instead, it may be due to nothing more than age. As we get older, the structure of our sleep undergoes subtle changes. The amount of time that adults spend each night in REM sleep begins to decline at around the age of forty. At that age, the brain begins a process of readjusting its sleep pattern and devoting more time to the lighter stages of sleep. Soon, the loud of a barking dog that someone was able to sleep through at the age of twenty-five is a nuisance that makes sleep impossible. These changes, a decade in the making, often become more apparent once someone turns fifty. By the time a person reaches sixty-five, he or she usually settles into a pattern marked by falling asleep around nine o'clock at night and waking up at three or four in the morning.

What many older adults call insomnia may in fact be an ancient survival mechanism. Carol Worthman, an anthropologist at Emory University in Atlanta, has argued that the modern comforts of silence, deep foam mattresses, and climate control have given us the expectation that sleep should always come easily. The wiring of our brains hasn't caught up with the comforts of our bedrooms, however. With no sharp claws or teeth to scare off potential predators, early humans were at their most defenseless when they laid down on the ground for several hours in the middle of the night.

Sleeping patterns that change as we age show that our brains expect us to be living and sleeping in a group, Worthman says. To illustrate this idea, she noted that the three basic stages of adulthood—teenage, middle age, old age—have drastically different sleep structures. Teenagers going through puberty find it impossible to fall asleep early and would naturally sleep past ten in the morning if given the choice. Their grandparents often fall asleep early in the night, but then find that they can't stay that way for more than three or four hours at a time.

Middle-aged adults typically fall between the middle of these two extremes, content to fall asleep early when circumstances allow it, yet able to pull an all-nighter when a work project calls for it. These overlapping shifts could be a way to ensure that someone in the family is always awake and keeping watch, or at least close to it. In this ancient system, it makes sense that older adults who are unable to move as fast as the rest of the family are naturally jumpy, never staying in deep sleep for long, simply because they were the most vulnerable to the unknown.

Those survival instincts are of little help when life takes place in a condo in Boca Raton. One 2003 poll by the National Sleep Foundation found that nearly seven out of every ten adults aged fifty-five to eighty-four reported frequent sleep problems. When sleeping pills or sleep apnea masks aren't the answer, what is left is perhaps the newest branch of sleep medicine: the science of how to naturally get a good night's rest.

12

Mr. Sandman

It started as a way to get a better grade. In the fall of 2003, Jason Donahue, then a junior at Brown University, sat listening in his living room to a friend who had just come back from a psychology lecture and was eager to share what he had learned. In class that day, his professor had discussed the concept of sleep inertia. Roughly speaking, the higher-functioning aspects of the brain—making decisions, recalling important facts, directing precise movements of the body—are compromised if a person is woken up during certain stages of sleep. Just like the physics of a moving object, the brain resists changes to its current state.

This phenomenon is most pronounced when the brain is

suddenly taken from deep, slow-wave sleep and faced with the complexities of navigating life. Logic becomes fuzzy, reactions slow down, and the brain typically wants nothing more than to fall back to sleep. The first scientists who noticed the condition called it sleep drunkenness. It often takes its most extreme form when a person is woken up from deep sleep during the first part of the night. In studies, subjects rustled from this stage of sleep were confused and disoriented by their surroundings, leading to strange behaviors like picking up a lamp next to the bed and talking into it as if it were a phone without realizing their error.

Sleep inertia is a well-known—and much-feared—consideration in aviation safety. Pilots who wake up suddenly from a cockpit nap are more likely to make poor decisions that cost lives. In May of 2010, Zlatko Glusica was the captain of an Air India Express plane carrying 166 passengers from Dubai to Mangalore, a bustling port city on India's southern coast. Among pilots, the city's airport is known for its short runway, making for a notoriously difficult landing. Glusica, fifty-three, was a veteran who had logged over ten thousand hours of flying time. More importantly, he had previously landed nineteen flights at Mangalore. Voice recordings recovered from the cockpit picked up the sound of Glusica snoring for much of the three-hour flight across the Arabian Sea. As the Boeing 737 approached the landing, Glusica woke up and took over the controls from his copilot. It was clear almost immediately that Glusica was in no condition to complete the task safely. His copilot warned him repeatedly that he was coming in at the wrong angle and that he should pull up and try again. But Glusica, groggy from sleep inertia, didn't process the red flags and went ahead with his flight path. The last sound on the cockpit recorder was the copilot screaming that they didn't have any runway left. The plane overshot its landing and burst into flames. Only eight people survived.

Back at Brown University, the college juniors had a very different set of considerations when it came to sleep inertia. Donahue recalled times when he had stayed up late studying for an exam, only to wake up the next morning foggy-headed and unable to concentrate for what seemed like hours. The information he knew the night before was lost and his test grades suffered. In his living room, Donahue began wondering whether it would be possible to time his sleep cycle perfectly so that he woke up at the ideal moment, effectively hacking his body's rhythms and tweaking them for his benefit. He began hanging out around Brown's School of Engineering, looking for someone who could develop a cheap and portable way to track the stages of sleep that wouldn't require the onerous wires and equipment found in professional sleep labs. That's when he met Ben Rubin. A fellow junior, Rubin was intrigued by the idea of capturing the unseen activity of the brain and using this information to improve his life.

With money won from a business competition, Donahue and Rubin set to work on a prototype of a device that would track the stages of sleep and wake a person up at an optimal moment in the sleep cycle. The concept they developed was relatively simple. It consisted of a brain wave–tracking monitor worn around the forehead during sleep that took a snapshot of the brain's activity every thirty seconds and estimated what stage of sleep a person was in. Before going to bed, a user would program what time he or she needed to be awake the next morning. The machine's alarm would go off up to a half hour before the deadline if there was a window of light sleep that allowed for a smooth transition into the new day. In the showdown between sleeping versus waking up, getting out of bed earlier won if it meant sidestepping sleep inertia.

Their first users, however, came away with an altered view of the product's function. As soon as they saw their total sleep time and the minute-by-minute account of their sleep cycles, they no longer cared about the alarm clock aspect of the device. "Our friends were the first testers, and some of them started freaking out when they saw that they were waking up eight times a night and couldn't remember it," Donahue told me. Tracking and analyzing the quality and quantity of sleep was more appealing than waking up at the perfect time. Donahue and Rubin went back to the drawing board and redesigned the product to focus on data collection.

Six years later, they launched the device as a consumer product. Now called the Zeo Personal Sleep Coach, it offered to unlock the mysterious hours of sleep to anyone who was willing to wear a black fabric headband that places a hard plastic square just above the eyebrows. Inside the square are three monitors that pick up electrical activity in the brain and lateral movements of the eyes. Together, they are on the look-out for the biological markers of sleep. When plugged into a base station that doubles as an alarm clock, the Zeo displays a graph that gives a precise breakdown of the prior night's sleep. The appearance of sleep spindles—two- to three-second bursts of voltage that rise and fall rapidly—on the graph hints that a person is at the shallowest portions of the roughly ninety-minute cycle. Long, slow waves signal the deepest sleep. And brain-wave activity resembling waking thought that comes at the same time as rapid movement of the eyes suggests that a person is dreaming. After wearing a Zeo headband all night, users can pinpoint that they were in dreaming sleep between 1:45 and 2:10, say, or that they woke up briefly at 3:30 and again an hour later.

But it is the second function of the Zeo that is revolutionary. Using an algorithm that takes into account the number of times a person wakes up, the time spent in deep sleep, and the total time spent sleeping, the Zeo rates each night of sleep and spits out a figure, known as a ZQ, that ranges from 0, for the worst sleep imaginable, to 120, the ideal sleep. With the advent of the ZQ rating, sleep entered the world of data and tracking. Implicit in any quantifiable scale comes the promise of improvement. Just like cholesterol levels, body weight, and blood pressure, sleep could now be translated into a numbered path that offered guideposts to personal betterment. Over time, a user can compare one night of sleep to another without resorting to intuition. "Measurement by itself isn't the end goal," Rubin told me. "The end goal is improvement, whether it's something that tracks running or sleeping or weight loss. Measurement allows you to do that in new ways."

Numbers are one way to fulfill an ambition that eludes nearly all of us. We each want perfect sleep. But getting there is much harder said than done. A thicket of studies have shown that humans, as a rule, do a terrible job of judging not only how they slept on any particular night but also what goes into making them sleep better. Accurately estimating how long it took to fall asleep and contrasting one night of sleep with another are two skills that are simply beyond our capacity. Sleep science, for much of its short history, offered little direction. With researchers understandably focused on disorders such as sleep apnea and parasomnias, the field of sleep medicine has spent more time researching why a night of sleep went wrong than why another went right.

Only recently has science figured out what goes into a good night of sleep. Falling asleep, and staying that way throughout the night, appears to be a battle with two fronts. The first takes place in the head. Between the time when a person lays his or her head on a pillow and the time when the brain sends out the first sleep spindles marking the onset of sleep, the mind must put aside its focus on its immediate surroundings and daily concerns. This process requires a person to give up direct control of his or her thoughts. At the same time, the body must be comfortable enough that the brain essentially forgets that they are attached. When something gets in the way of either, the end result is often insomnia.

These dual tracks toward good sleep are rarely recognized by the average person. Instead, many assume that physical comfort is the only thing standing between them and a good night's rest. Mattresses, for this reason, are perhaps the most basic consumer purchase and the least understood. Many people who walk into a mattress store have only vague notions of what they think they will like. It's no surprise, then, that the number of questions *Consumer Reports* fields about purchasing a new bed ranks only behind the number concerning buying a new car.

The biggest question—whether a bed should be hard or soft—has a long and confusing history. In an 1833 article in an Irish newspaper called the *Dublin Penny Journal*, for instance, a writer known as the Celebrated Doctor Abercrombie suggested that "the mattress, or bed, on which we lie, ought to always be rather hard. Nothing is more injurious to health than soft beds; they effeminate the individual, render his flesh soft and flabby, and incapacitate him from undergoing any privation." In the 1970s and 1980s, a rebellion against firm mattresses accounted for the brief popularity of waterbeds. Sales of these heavy, leak-prone mattresses peaked at 3.8 million in 1988 and then plummeted. Waterbeds have since been rebranded as flotation mattresses, though the name change hasn't done much to revive sales. In 2008, the medical journal *Spine* seemed to settle the question of firmness. It found that there was little difference in back pain between those who slept on hard mattresses and those who slept on softer ones. How hard a person likes his or her bed is a personal preference and nothing more.

In fact, the bed that you find the most comfortable will most likely be the one that you are already sleeping on. This inclination toward the routine was first noticed in a research study conducted in the 1950s. In it, subjects were asked to rate the firmness of their mattress at home and their overall sleep quality. They then slept on three mattresses in a lab—one hard, one soft, and one in-between—over three separate nights. At the end of the study, researchers matched overall satisfaction with each type of mattress. The biggest factor that influenced the rankings was how closely each bed matched the one that a subject had at home. Fifty years later, researchers at a German hospital again decided to search for the perfect mattress. With a half century of research into sleep at their side, they reasoned that they could identify the ideal firmness that would allow the maximum level of comfort for every patient in the hospital. But the human body wasn't as efficient as they had hoped. "There was no global preference for any of the beds tested," the German team found. They continued, with a pang of resignation. "Each individual seemed to develop its own sleep pattern which meant that results could be only compared for the same subject." Just as in the study from the 1950s, patients liked what was the most familiar.

A comfortable surface isn't always necessary for quality sleep. In the late 1960s, William Dement—one of the premier researchers in the history of sleep science who you may remember from an earlier chapter—received a commission from a company that had recently built a prototype of a high-tech mattress. Inside of it, warm air flowed through billions of tiny, ceramic beads. The end result was the feeling of a cushion built out of heated mud. "Everyone in our lab agreed that it was the most comfortable bed they had ever lain on," Dement later noted. The company that hired him asked Dement to determine how well people slept on its product, which was expected to retail for several-thousand dollars, compared with a conventional mattress. To make the final result more dramatic, Dement decided to include a third option in his study: sleeping on a concrete floor with no padding. Volunteers gamely spent a night on each of the three, and Dement's team later evaluated the results. "We were absolutely flabbergasted," he wrote. There were no significant differences in the quality of a volunteer's sleep or the total number of hours spent sleeping on the three surfaces. Given the option between a concrete floor and a high-tech mattress, subjects in the study experienced roughly the same night of sleep either way.

While a comfortable mattress may have little impact when it comes to sleep quality, there are several other aspects of the bedroom that do. Taken together, they form what specialists call sleep hygiene. Most are common sense. It is obviously not a good idea to drink coffee in the evening if it keeps you up at night. Nor is drinking alcohol before bedtime a smart move. Alcohol may help speed the onset of sleep, but it begins to take its toll during the second half of the night. As the body breaks down the liquid, the alcohol in the bloodstream often leads to an increase in the number of times a person briefly wakes up. This continues until the blood alcohol level returns to zero, thereby preventing the body from getting a full, deep, restorative sleep.

Developing a few habits with the circadian rhythm in mind will most likely make sleep easier. Adequate exposure to natural light, for instance, will help keep the body's clock in sync with the day-night cycle and prime the brain to increase the level of melatonin in the bloodstream, which will then bring on sleepiness around ten o'clock each night. By the same token, bright lights—including the blue-and-white light that comes from a computer monitor or a television screen—can deceive the brain, which registers it as daylight. Lying in bed watching a movie on an iPad may be relaxing, but the constant bright light from the screen can make it more difficult for some people to fall asleep afterward. Other common suggestions from sleep doctors include maintaining a consistent bedtime, using the bedroom only for sex or sleeping, and turning the lights down low in the home about a half hour before climbing into bed.

Recent studies have shown that body temperature also plays an outsized role in getting decent sleep. In addition to the appearance of brain waves like sleep spindles, one of the biological markers of the onset of sleep is a drop in core body temperature. At the same time, the temperature of the feet and hands increases as the body gives off heat through its periphery, which explains why some people like to have their feet sticking out of the covers as they fall asleep. The body's tendency to release heat during the night is one reason why some mattresses are said to be uncomfortable—they "sleep hot." In the simplest explanation, the fabric and materials that make up some beds trap the heat the body is releasing. The result can make the bed feel like an oven, preventing the body from cooling itself down. The core temperature of the body falls naturally for most of us after ten o'clock each night as a result of the circadian rhythm. When this doesn't happen, chronic insomnia is often a result. Researchers at an Australian university found that patients with insomnia had significantly higher core body temperatures when attempting to fall asleep than those who identified themselves as good sleepers.

Assisting the body in its cooling process, then, is a natural way to improve sleep. One study by researchers in Lille, a city in northeastern France, found that subjects fell asleep faster and had a better overall quality of sleep following behaviors that cooled the body, such as taking a cold shower right before bed. The best predictor of quality sleep was maintaining a room temperature in a narrow band between 60 and 66 degrees Fahrenheit (or 16 to 19 degrees Celsius). Temperatures above or below this range often led subjects to become restless, either tossing or turning from being too hot or shivering from the cold. The study assumed, of course, that the subjects were sleeping in pajamas with at least one sheet covering them. Being tenacious researchers, the French team also investigated the proper room temperature for those who prefer to sleep naked. They found that it was a much higher range, 86 to 90 degrees Fahrenheit, or 30 to 32 degrees Celsius.

Lowering the body's temperature is not the only physical activity that can make the night easier. Even a small increase in the amount of exercise a person gets leads to measurable improvements in the time that it takes to fall asleep and stay that way. This is particularly true for older adults. In one study of men and women with clinical depression, undergoing ten weeks of weight training resulted in significant improvements in sleep quality. Another study of sedentary adults with a history of sleep problems found that a four-month exercise regiment dramatically reduced the time it took them to fall asleep. And in one of the most intriguing studies, researchers in Seattle tracked nearly two hundred overweight or obese women over the age of fifty for one year. At the beginning of the study, each woman averaged less than an hour of moderate to vigorous exercise each week. One group of subjects agreed to stick with an exercise program for the next twelve months, and the rest maintained their normal lifestyle. A year later, it wasn't surprising that those who exercised reported a better quality of sleep than those who remained sedentary. But there were also significant differences in sleep quality among the women who exercised. Those who spent more than three and a half hours involved in physical activities each week had less trouble falling asleep than those who exercised for less than three hours.

At first glance, it appears obvious that exercise's benefit for sleep is simply a matter of physical exhaustion. After all, those women who slept the best were also the ones who were the most active. Maybe their bodies were so depleted that they just needed more sleep. But the relationship between sleep and exercise is yet another instance in which the brain isn't as straightforward as it seems.

What may matter more than the amount of time spent doing strenuous physical activity is how hard the brain considers the work to be. One study, completed by Swiss researchers and published in the *Journal of the American College of Sports Medicine,* focused on nearly nine hundred college students in Switzerland. Each person in the study tracked how much he or she exercised during the week. Subjects were also asked to fill out two questionnaires. In one, they rated how well they slept, on a scale of 1 to 10. In the other, they rated their overall fitness level, again on a scale of 1 to 10. Overall, the research team found no link between how many hours a student spent working out and how well he or she slept each night.

The self-assessments uncovered other surprising results, however. Nearly a fifth of those who rated themselves low on the fitness scale were, in reality, among the most physically active participants in the entire study group. These subjects were working out all the time, but they simply felt like they weren't doing enough. Those perceptions carried over into their sleep. Although they were more active than other members of the study, subjects in this group reported sleep qualities that were below average. They were doing the work but not getting the rewards.

This followed the findings of other studies which discovered that the connection between exercise and sleep isn't just a matter of tiring oneself out. One, for instance, tracked college students for a little longer than three consecutive months. At the end of this time span, researchers went back and identified the eleven days on which subjects were the most physically active and the eleven days with the lowest amount of movement. They then compared a subject's quality of sleep on the nights after he or she was the most physically taxed with the quality on the nights after he or she presumably had energy to spare. Logic would hold that a busier day means a better sleep. But there was little difference. The study found no relationship between an individual activity level during the day and the quality of sleep that night, which means that a long run alone won't provide the answer to a better night's sleep.

But the same study did uncover one of the brain's quirks. Intriguingly, the subjects who thought they were in good shape slept well—even if they weren't actually exercising as much as others in the study. It appeared that whatever amount of time these subjects spent working out was enough to cross the mental threshold that told them not to worry about their fitness level. As they laid down each night, their lack of concern about whether they were exercising enough gave them one less thing that could stand in the way of drifting off to sleep. Their minds thought their bodies were meeting the standard, and they acted accordingly. As the lead researcher for the Swiss study told the *New York Times,* "What people think is more important than what they do."

This same principle applies to other tactics to improve sleep that could fall under the general category of holistic medicine. Yoga, acupuncture, and massage have all been linked with improved sleep, in part because they put both the body and the mind at ease. The mental aspect of each activity can't be overlooked. In one study, patients were asked to do a form of breathing exercises that would be familiar to anyone who has finished a yoga class in corpse pose. As subjects laid on their back with their eyes closed each night, they were instructed to focus on their breathing by thinking the word *in* every time they inhaled and *out* with each exhalation. The technique proved as effective as other forms of relaxation strategies used to treat insomnia.

The secret of the Zeo, for all of the technology that goes into capturing a person's brain waves and translating them into a number that rates each night of sleep, may be that it accomplishes the same thing as these mental techniques. By providing feedback in the form of an easily understood scale, the device could simply be giving its users the same sense of satisfaction that those lucky Swiss college students felt. They weren't among the most physically active in a literal sense, but reality didn't matter. Their minds believed that they were fit, so therefore they acted—and slept—accordingly. The secret to a good night's rest could just be stopping the mind from getting in its own way.

13

Good Night

When I began this book, I did so with a selfish plan. By interviewing experts specializing in every aspect of sleep science, I expected to come away from the project with all of my own sleepwalking problems solved.

It didn't quite work out that way. This was never more obvious than on a hot July afternoon I spent in San Antonio. I was there for the largest annual meeting in the country of the doctors, researchers, and academics who make the study of sleep their lives. There, spread out across a convention floor the size of four football fields, were booths selling every imaginable product that could possibly affect a person's sleep. Some vendors sold T-shirts with pockets of air stitched into the back, claiming they prevented snoring by forcing a person to sleep on his or her side. Beside them were floor-to-ceiling booths from companies marketing pills for patients with narcolepsy and other disorders. One floor display was so large there was enough room for a chef wearing a tall white hat to bake warm peanut butter cookies in the middle of it. And in front of me, wearing a bright red shirt, was a man named Mike.

Mike's product looked to be nothing more than a big glass jar on top of a record player. Inside the jar was a lifelike plastic rat with wires attached to its temples. If this had been a real rat, then it wouldn't have liked its predicament. That's because the entire purpose of the machine was to keep the animal awake for long stretches at a time without any human labor. Apparently, keeping a rat from falling asleep is harder than it looks. "It's very labor-intensive," Mike said, and I don't doubt him. "If you're a university, you're relying on grad students having to poke them. They've got to keep that up for a day or two, maybe three if they really want to finish their thesis."

Mike worked for a company in Kansas called Pinnacle Technology. For $7,500, I could have purchased the 8400-K1-Bio machine that stood in front of me. The selling point of the system, Mike explained, was that it was guaranteed to keep a mouse or rat awake without painful electric shocks, which could corrupt the data of sleep deprivation experiments. Mike pointed out its features. If the rat fell asleep, the tiny neurotransmitter wires on its head would recognize it in an instant. The machine would then spring to action. In less than a second, the heavy plastic rod that was resting against the plastic rat's feet would start whirling around, and keep at it until the rat woke up. No poking required.

It was a product that said more about sleep than it intended. As I looked at the machine, a realization arrived fully formed in my head: the more you know about sleep, the more its strangeness unnerves you. Before that moment, I would never have conceived of a product whose sole purpose was to deprive rats of sleep. And before I started this book, I wouldn't have thought that sleepwalkers can kill, or that companies hope millions of people in China consume fast food and develop obesity-related sleeping disorders, or that one of the most popular prescription drugs works by making it harder for someone to form memories. Even after spending hundreds of hours with experts in various fields and reading a tall stack of research reports, I still saw sleep as this strange part of life that was all the more mysterious because of its importance. The existence of Mike's machine suggested that there were still more puzzling aspects of sleep to be found, more researchers designing odd studies searching for the purpose of sleep, and more outcomes than one would think possible.

There was an upside to my newfound knowledge of this strange world, however. By that point, I knew enough about sleep that I began to improve my own. Unlike most people with a sleep problem, insomnia had never been an issue for me. Instead, my troublesome nights were, and are, those that I spend kicking, talking, or at their worst, walking down the hallway while still dreaming.

My sleep improvement plan was relatively simple. The first step consisted of using a Zeo, the brain wave–tracking device, for a month. Like the first testers of the product, I too was drawn to the fact that I could see in front of me a record of all of my previous nights of sleep, from the times that I woke up to the hours I spent dreaming. On the first night that I used the device, my ZQ rating was a 40—or about a third of the way on the scale in which 120 was the perfect night of sleep. I wasn't surprised by the low number, considering the fact that I was attempting to sleep with a brain-wave monitor on my head. I had slept poorly during the night I spent in the professional sleep lab, too. The next night the headband felt slightly more natural. When I woke up, I sensed that I had had a normal night of sleep. My ZQ rating jumped to a 68. The machine revealed that I had woken up several times during the night that I couldn't remember, but by that time that fact had ceased to be alarming. I also had spent what seemed like an adequate amount of time in dreaming sleep. That morning my wife said that she vaguely remembered hearing me talk in my sleep, but nothing out of the ordinary.

Still, I wanted to boost my ZQ up to at least 100. Maybe doing so would recalibrate my sense of what a good night of sleep really was. I started following all of the advice that I had received over the months. I began eating breakfast in the sunniest corner of our apartment every morning in order to improve my body's synchronization with the day-night cycle. I became a fixture at yoga classes at the gym. And I roamed around the apartment turning lights off a half hour before I went to sleep. Each night,

I continued to place the Zeo headband on and track my results. My ZQ score climbed steadily, rising from 68 one night to a 74 the next and all the way up to 88.

It peaked at a 94. I never made it to my self-imposed goal of a 100, but I was comforted by the fact that every other measure of my sleep improved. I woke up feeling more refreshed than normal. I had an easier time remembering things like where I put my keys and the date and time of my next dentist appointment. And the subway commute into and out of midtown Manhattan every workday felt calmer. Most important, I began to get a better sense of my body when it came to sleep. After one particularly stressful day at work, I felt an odd sensation as I got ready for bed: for some reason, I could tell that I was going to have a rough night sleep talking or kicking. Instead of fighting this fact, or worse yet, ignoring it, I understood enough about sleep to realize there was little I could do to prevent it on this night. I decided to sleep on the couch.

At first glance, it doesn't seem that sleeping better was all that life-changing. After all, there is no guarantee that I won't sleepwalk again into one of my hallway walls, or, even worse, sleepwalk into something more painful. I may sleepwalk again tonight, two Tuesdays from now, or perhaps never again. That's just another part of the puzzle of sleep. But, though its effects were subtle, devoting extra time and attention to this most basic of human needs impacted nearly every minute of my day. Because I was improving my sleep, I was improving my life. And all it took was treating sleep with the same respect that I already gave other aspects of my health. Just as I wouldn't eat a plate of chili-cheese fries every day and expect to continue to fit into my pants, I structured my life around the idea that I couldn't get only a few hours of sleep and expect to function properly. If there was one thing that I took away from my talks with experts more than any other, it is that getting a good night's sleep takes work.

And that work is worth it. Health, sex, relationships, creativity, memories—all of these things that make us who we are depend on the hours we spend each night with our heads on the pillow. By ignoring something that every animal requires, we are left turning to pills that we may not need, experiencing health problems that could be tamed, and pushing our children into sleep-deprived lives that make the already tough years of adolescence more difficult. And yet sleep continues to be forgotten, overlooked, and postponed. Any step—whether it comes in the form of exercise, therapy, or simply reading a book like this one—that helps us to realize the importance of sleep inevitably pushes us toward a better, stronger, and more creative life.

Sleep, in short, makes us the people we want to be. All you have to do is close your eyes.

Acknowledgments

This is a book that came about because I hurt myself. Everything that has happened to me since that night suggests that sleepwalking can be one of life's happy accidents.

I have been extremely fortunate to work with a team of smart, dedicated people, starting with my agent pair of Larry Weissman and his wife, Sascha Alper. Together, they calmly guided me through a process that took us from shaping the proposal over drinks in their living room to talking through the edits of the final manuscript. Along the way, they never wavered in their enthusiasm for the project. While we were walking in Park Slope one night, Larry gave me just the advice that a fretting first-time author needs: "Dave, remember that it's a marathon." With that, every writing or research problem seemed manageable.

Jill Bialosky, my editor at Norton, was another godsend. She took the first draft of the manuscript and deftly identified what needed to be expanded, what needed to be fixed, and what needed to be shucked. An accomplished poet and author, Jill is exactly the sort of editor one dreams about when fermenting the crazy notion of moving to New York and writing a book. This book is immeasurably better because of her work. Her assistant, Alison Liss, provided a smart second edit as well, giving me the benefit of two editors for the price of one.

I'm grateful to the rest of the group at Norton who put time into this project. Eleen Cheung developed a beautiful cover. Mary Babcock, as copyeditor, gracefully honed my prose and saved me from many embarrassing errors. Any mistakes that remain are mine alone.

I knocked on many doors while conducting research for this book, and was constantly surprised at how willing people were to answer. The men and women I interviewed were generous with their time, patiently explaining concepts ranging from neuroscience to the history of furniture. If only every reporter was as lucky.

I wouldn't be in a position to write an acknowledgments section in the first place if it wasn't for friends and mentors along the way. Matthew Craft, Joyce Macek, Zack O'Malley Greenburg, Jon Bruner, Asher Hawkins, Tim Stelloh, Alan Yang, Dirk Smillie, Jonathan Fahey, Michelle Conlin, and Laurie Burkitt gave advice, contributed ideas, offered reporting assistance, or took the time to listen to me ramble about whatever I was working on at the moment. Mary Ellen Egan, Neil Weinberg, Larry Reibstein, Kevin Shinkle, and Jennifer Merritt went out of their way to make it possible for me to write a book while holding down a full-time job.

I come from a family in which nearly everyone works in education, so I would be committing an unforgivable sin to neglect the teachers who shaped me along the way. Robert Ayres, my high school journalism teacher, was demanding, opinionated and gruff, and I am a better writer and thinker because of him. William Serrin, Brooke Kroeger, Robert Boynton, Craig Wolff, and Michael Norman made the NYU journalism program one of the best investments I have made. And Diego Ribadeneira, my editor at the *New York Times*, taught me how to find a compelling story in the most obscure places. It is because of him that I can proudly say that I have written about unicyclists, dog parks, and people who fall asleep in museums.

Finally, I have to give an enormous round of thank-yous to my family. Anthony and Maryanne Petrizio, Robert and Gina Scott, and Ryan Randall each provided invaluable kindness and encouragement. My parents, Kenneth and Diane Randall, started me on the path to becoming a writer by maintaining a house where a pile of books could be found in every corner. I will always be thankful for their love and support.

And, of course, the biggest thank-you goes to my wife and best friend, Megan, who read every single word that made it into this book countless times (and many, many more words that didn't). She was a constant source of support and insight who made this project possible. And she accomplished this with a sweet-natured charm, all while sleeping next to someone who routinely kicks her in the middle of the night. That truly is more than anyone could ask for.

Bibliography

Chapter 1: I Know What You Did Last Night

Basner, Mathias, and David F. Dinges. "Dubious Bargain: Trading Sleep for Leno and Letterman." *Sleep*, vol. 32 (June 2009).

Dement, William C., and Christopher Vaughan. *The Promise of Sleep: A Pioneer in Sleep Medicine Explores the Vital Connection between Health, Happiness, and a Good Night's Sleep.* New York: Delacorte Press, 1999.

Dreifus, Claudia. "Eyes Wide Shut: Thoughts on Sleep." *New York Times*, October 23, 2007.

Everson, C A., B. M. Bergmann, and A. Rechtschaffen. "Sleep Deprivation in the Rat: III. Total Sleep Deprivation." *Sleep*, vol. 12 (February 1989).

Gillin, J. Christian. "How Long Can Humans Stay Awake?" *Scientific American*, March 25, 2002.

Max, D. T. *The Family That Couldn't Sleep*. New York: Random House, 2006.

Palmer, Brian. "Can You Die from a Lack of Sleep?" *Slate*, May 11, 2009.

Pressman, Mark R. "Sleepwalking Déjà Vu." *Sleep*, vol. 32 (December 2009).

Rattenborg, N. C., S. L. Lima, and C. J. Amlaner. "Half-Awake to the Risk of Predation." *Nature*, vol. 397 (February 4, 1999).

Stickgold, Robert. "Neuroscience: A Memory Boost While You Sleep." *Nature*, vol. 444 (November 30, 2006).

Vyazovskiy, Vladyslav V., Umberto Olcese, Erin C. Hanlon, Yuval Nir, Chiara Cirelli, and Giulio Tononi. "Local Sleep in Awake Rats." *Nature*, vol. 472 (April 28, 2011).

Chapter 2: Light My Fire

Akerstedt, T., and M. Gillberg. "A Dose-Response Study of Sleep Loss and Spontaneous Sleep Termination." *Psychophysiology*, vol. 23 (May 1986).

Arimura, M. "Sleep, Mental Health Status, and Medical Errors among Hospital Nurses in Japan." *Industrial Health*, vol. 48 (November 2010).

Barger, Laura K., Brian E. Cade, Najib T. Ayas, John W. Cronin, Bernard Rosner, Frank E. Speizer, and Charles A. Czeisler. "Extended Work Shifts and the Risk of Motor Vehicle Crashes among Interns." *New England Journal of Medicine*, vol. 352 (January 13, 2005).

Chepesiuk, R. "Missing the Dark: Health Effects of Light Pollution." *Environmental Health Perspectives*, vol. 117 (January 2009).

Ekirch, A. Roger. *At Day's Close: Night in Times Past*. New York: W. W. Norton, 2005.

Fox, Karen. "Sleeping the Sleep of Our Ancestors." *Science*, vol. 262, no. 5137 (November 19, 1993).

Goodman, Al. "Snoring to Success in Spain's First National Siesta Championship." *CNN World*, October 15, 2010.

Hathaway, Warren E. "Effects of School Lighting on Physical Development and School Performance." *Journal of Educational Research*, vol. 88 (March/April 1995).

McLean, Renwick. "For Many in Spain, Siesta Ends." *New York Times*, January 1, 2006.

Ohayon, M. M., M. H. Smolensky, and T. Roth. "Consequences of Shiftworking on Sleep Duration, Sleepiness, and Sleep Attacks." *Chronobiology International*, vol. 27 (May 2010).

Stross, Randall E. *The Wizard of Menlo Park: How Thomas Alva Edison Invented the Modern World.* New York: Crown, 2007.

U.S. Chemical Safety and Hazard Investigation Board. *Investigation Report: Refinery Explosion and Fire.* Report no. 2005-04-I-TX. March 2007.

Wehr, Thomas A. "In Short Photoperiods, Human Sleep Is Biphasic." *Journal of Sleep Research,* vol. 1 (June 1992).

Chapter 3: Between the Sheets

"Bed Sharing 'Bad for Your Health.'" BBC News, September 9, 2009.

Coontz, Stephanie. *Marriage: A History from Obedience to Intimacy, or How Love Conquered Marriage.* New York: Viking Adult, 2005.

Halliday, Stephen. "Death and Miasma in Victorian London: An Obstinate Belief." *British Medical Journal,* vol. 323 (December 22, 2001): 1469–1471.

Hinds, Hilary. "Together and Apart: Twin Beds, Domestic Hygiene and Modern Marriage, 1890–1945." *Journal of Design History,* vol. 23, no. 3 (2010).

Meadows, Robert. "The 'Negotiated Night': An Embodied Conceptual Framework for the Sociological Study of Sleep." *Sociological Review,* vol. 53, no. 2 (May 2005): 240–254.

Mondello, Bob. "Remembering Hollywood's Hays Code, 40 Years On." *All Things Considered,* NPR, August 12, 2008.

Rosenblatt, Paul C. *Two in a Bed: The Social System of Couple Bed Sharing.* Albany: State University of New York Press, 2006.

Rozhon, Tracie. "To Have, Hold and Cherish, until Bedtime." *New York Times,* March 11, 2007.

Troxel, W. M. "It's More than Sex: Exploring the Dyadic Nature of Sleep and Its Implications for Health." *Psychosomatic Medicine,* vol. 72, no. 6 (July/August 2010).

Troxel, W. M., D. J. Buysse, M. Hall, and K. A. Matthews. "Marital Happiness and Sleep Disturbances in a Multi-Ethnic Sample of Middle-Aged Women." *Behavioral Sleep Medicine,* vol. 7, no. 1 (2009).

Troxel, W. M., T. Robles, M. Hall, and D. J. Buysse. "Marital Quality and the Marital Bed: Examining the Covariation between Relationship Quality and Sleep." *Sleep Medicine Reviews,* vol. 11 (October 2007).

Weiner, Stacy. "Estranged Bedfellows." *Washington Post,* January 10, 2006.

Chapter 4: And Baby Makes Three

Blum, David. "When Lullabies Aren't Enough: Richard Ferber." *New York Times Magazine,* October 9, 1994.

Brown, Charity M. "Women Are More Likely Than Men to Give Up Sleep to Care for Children and Others." *Washington Post,* February 14, 2011.

Burgard, Sarah. "The Needs of Others: Gender and Sleep Interruptions for Caregivers." *Social Forces,* vol. 89, no. 4 (June 2011).

Ferber, Richard. *Solve Your Child's Sleep Problems.* New York: Fireside, 1986.

Gomez, Mark. "Debate Rages over Having Babies Sleep with Parents." *San Jose Mercury News,* July 4, 2010.

Huang, Xiao-na. "Co-sleeping and Children's Sleep in China." *Biological Rhythm Research,* vol. 41, no. 3 (2010).

McKenna, James J., Helen L. Ball, and Lee T. Gettler. "Mother–Infant Cosleeping, Breastfeeding and Sudden Infant Death Syndrome: What Biological Anthropology Has Discovered about Normal Infant Sleep and Pediatric Sleep Medicine." *American Journal of Physical Anthropology,* vol. 134 (November 2007).

Meltzer, Lisa J., and Jodi A. Mindell. "Impact of a Child's Chronic Illness on Maternal Sleep and Daytime Functioning." *Archives of Internal Medicine,* vol. 166 (September 18, 2006).

Meltzer, Lisa J., and Jodi A. Mindell. "Relationship between Child Sleep Disturbances and Maternal Sleep, Mood, and Parenting Stress: A Pilot Study." *Journal of Family Psychology,* vol. 21 (March 2007).

Mindell, J. A., A. Sadeh, J. Kohyama, and T. H. How. "Parental Behaviors and Sleep Outcomes in Infants and Toddlers: A Cross-Cultural Comparison." *Sleep Medicine*, vol. 11 (April 2010).

Mindell, Jodi A., Lorena S. Telofski, Benjamin Wiegand, and Ellen S. Kurtz. "A Nightly Bedtime Routine: Impact on Sleep in Young Children and Maternal Mood." *Sleep*, vol. 32 (May 2009).

Rudd, Matt. "Move over, Darling—Preferably Right into the Other Bedroom: A Study Says the Best Way for a Couple to Get a Good Night's Rest Is to Sleep Apart." *Sunday Times* (London), September 13, 2009.

Seabrook, John. "Sleeping with the Baby." *New Yorker*, November 8, 1999.

Sobralske, Mary C. "Risks and Benefits of Parent/Child Bed Sharing." *Journal of the American Academy of Nurse Practitioners*, vol. 21 (September 2009).

Solter, Aletha. "Crying for Comfort: Distressed Babies Need to Be Held." *Mothering*, no. 122 (January/February 2004).

Stearns, Peter N., Perrin Rowland, and Lori Giarnella. "Children's Sleep: Sketching Historical Change." *Journal of Social History*, vol. 30 (Winter 1996).

Weissbluth, Marc. *Happy Sleep Habits, Happy Child*. New York: Ballantine Books, 1987.

Chapter 5: What Dreams May Come

Barrett, Deidre, and Patrick McNamara, eds. *The New Science of Dreaming*. Vol. 3: *Cultural and Theoretical Perspectives*. Westport, CT: Praeger, 2007.

Berlin, K. L. "Nightmare Reduction in a Vietnam Veteran Using Imagery Rehearsal Therapy." *Journal of Clinical Sleep Medicine*, vol. 6 (October 2010).

Blagrove, M., J. Henley-Einion, A. Barnett, D. Edwards, and C. Heidi Seage. "A Replication of the 5–7 Day Dream-Lag Effect with Comparison of Dreams to Future Events as Control for Baseline Matching." *Consciousness and Cognition*, vol. 20, no. 2 (June 2010).

Dement, W. C. "Recent Studies on the Biological Role of Rapid Eye Movement Sleep." *American Journal of Psychiatry,* vol. 122, no. 4 (October 1965).

Dement, William C., and Christopher Vaughan. *The Promise of Sleep: A Pioneer in Sleep Medicine Explores the Vital Connection between Health, Happiness, and a Good Night's Sleep.* New York: Delacorte Press, 1999.

Dixit, Jay. "Dreams: Night School." *Psychology Today,* vol. 40, no. 5 (November/December 1, 2007).

Empson, Jacob. *Sleep and Dreaming.* New York: Harvester Wheatsheaf, 1993.

Freud, Sigmund. *The Interpretation of Dreams.* Joyce Crick, trans. New York: Oxford University Press, 1999.

Gottesmann, Claude. "Discovery of the Dreaming Sleep Stage: A Recollection." *Sleep,* vol. 32 (January 2009).

Hall, C. S. "A Cognitive Theory of Dream Symbols." *Journal of General Psychology,* vol. 48 (1953).

Jouvet, M. "Paradoxical Sleep: A Study of Its Nature and Mechanisms," in K. B. Akert, C. Bally, J. P. Schadé, eds. *Sleep Mechanisms. Progress in Brain Research,* vol. 18. Amsterdam: Elsevier, 1965.

Mautner, B. "Freud's Irma Dream: A Psychoanalytic Interpretation." *International Journal of Psychoanalysis* (February 1991).

Murphy, Kate. "Take a Look inside My Dream." *New York Times,* July 9, 2010.

Nielsen, T. A., D. Kuiken, G. Alain, P. Stenstrom, and R. A. Powell. "Immediate and Delayed Incorporations of Events into Dreams: Further Replication and Implications for Dream Function." *Journal of Sleep Research,* vol. 13, no. 4 (December 2004).

Pick, Daniel, and Lyndal Roper, eds. *Dreams and History: The Interpretation of Dreams from Ancient Greece to Modern Psychoanalysis.* New York: Brunner-Routledge, 2004.

Reed, Charles F., Irving E. Alexander, and Silvan S. Tomkins, eds. *Psychopathology: A Source Book.* Boston: Harvard University Press, 1958.

Rock, Andrea. *The Mind at Night: The New Science of How and Why We Dream.* New York: Basic Books, 2004.

Chapter 6: Sleep on It

Callaway, Ewen. "Dreams of Doom Help Gamers Learn: The Dreams of Video Game Players Suggest That Nocturnal Visions Have a Practical Role: Helping Us to Learn New Skills." *New Scientist*, vol. 15 (November 2009).

Dement, William C. *Some Must Watch while Some Must Sleep*. San Francisco: San Francisco Book Company, 1976.

Durrant, S. J., C. Taylor, S. Cairney, and P. A. Lewis. "Sleep-Dependent Consolidation of Statistical Learning." *Neuropsychologia*, vol. 49 (April 2011).

Galenson, David. "Innovators: Songwriters." NBER Working Paper no. 15511. Cambridge, MA: National Bureau of Economic Research, November 2009.

Hoffman, Jascha. "Napping Gets a Nod at the Workplace." *BusinessWeek*, August 26, 2010.

Horne, Jim. *Sleepfaring: A Journey through the Science of Sleep*. Oxford: Oxford University Press, 2006.

Louie, K., and M. A. Wilson. "Temporally Structured REM Sleep Replay of Awake Hippocampal Ensemble Activity." *Neuron*, vol. 29 (January 2001).

Mednick, Sarnoff A. "The Associative Basis of the Creative Process." *Psychological Review*, vol. 69, no. 3 (1962).

Mednick, S. C., S. P. A. Drummond, G. M. Boynton, E. Awh, and J. Serences. "Sleep-Dependent Learning and Practice-Dependent Deterioration on an Orientation Discrimination Task." *Behavioral Neuroscience*, vol. 122 (April 2008).

Mednick, S. C., J. Kanady, D. Cai, and S. P. A. Drummond. "Comparing the Benefits of Caffeine, Naps and Placebo on Verbal, Motor, and Perceptual Memory." *Behavioral Brain Research*, vol. 3 (November 2008).

Mollicone, D. J., H. Van Dongen, and D. F. Dinges. "Optimizing Sleep/Wake Schedules in Space: Sleep during Chronic Nocturnal Sleep Restriction with and without Diurnal Naps." *Acta Astronautica*, vol. 60 (February–April 2007).

Moorcroft, William H. *Sleep, Dreaming and Sleep Disorders: An Introduction.* Laham, MD: University Press of America, 1993.

Pierre, Maquet, and Ruby Perrine. "Insight and the Sleep Committee." *Nature*, vol. 427 (January 22, 2004).

Povich, Shirley. "The 1964 U.S. Open: Victory in the Heat of Battle." *Washington Post,* June 11, 1997.

Stickgold, Robert. "A Few Minutes of Shut-Eye at Work Could Be Good for Business." *Harvard Business Review*, vol. 87, no. 10 (2009).

Stickgold, Robert, April Malia, Denise Maguire, David Roddenberry, and Margaret O'Connor. "Replaying the Game: Hypnagogic Images in Normals and Amnesics." *Science*, vol. 13 (October 2000).

Tucker, Matthew A., and William Fishbein. "Enhancement of Declarative Memory Performance following a Daytime Nap Is Contingent on Strength of Initial Task Acquisition." *Sleep*, Vol. 31 (February 2008).

Tupper, Fred. "Lema Takes British Open Golf with 279, Beating Nicklaus by Five Strokes." *New York Times*, July 11, 1964.

Wagner, U., S. Gais, H. Haider, R. Verleger, and J. Born. "Sleep Inspires Insight." *Nature*, vol. 427 (January 2, 2004).

Walker, Matthew P. "Sleep to Remember." *American Scientist*, vol. 94, no. 4 (July/August 2006).

Walker, Matthew P., Tiffany Brakefield, Alexandra Morgan, J. Allan Hobson, and Robert Stickgold. "Practice with Sleep Makes Perfect: Sleep-Dependent Motor Skill Learning." *Neuron*, vol. 35 (July 3, 2002).

Wilson, M. A. "Hippocampal Memory Formation, Plasticity, and the Role of Sleep." *Neurobiology of Learning and Memory*, vol. 78 (November 2002).

Chapter 7: The Weapon "Z"

Armstrong, Benjamin. "Are We Driving the Ship Drunk?" *Proceedings* (U.S. Naval Institute), vol. 136, no. 2 (February 2010).

Balkin, Thomas. "Managing Sleep and Alertness to Sustain Performance in the Operational Environment." Presentation notes, NATO, Nevilly-sur-Seine, France, 2005.

Berthoz, Alian. *Emotion and Reason: The Cognitive Neuroscience of Decision Making.* Oxford: Oxford University Press, 2003.

Committee on Military Nutrition Research, Food and Nutrition Board, Institute of Medicine. *Caffeine for the Sustainment of Mental Task Performance: Formulations for Military Operations.* Washington, D.C.: National Academy Press, 2001.

Driskell, James E., and Brian Mullen. "The Efficacy of Naps as a Fatigue Countermeasure: A Meta-Analytic Integration" *Human Factors,* vol. 47, no. 2 (Summer 2005).

Halbfinger, David M. "Hearing Starts in Bombing Error That Killed 4." *New York Times,* January 15, 2003.

Harrison, Yvonne, and James Horne. "The Impact of Sleep Deprivation on Decision Making: A Review." *Journal of Experimental Psychology: Applied,* vol. 6, no. 3 (2000).

"Information Paper: DARPA's Preventing Sleep Deprivation Program." Published on DARPA website (www.dtic.mil/cgi-bin/GetTRD?AD=ADA521349&Location=U2&doc=GetTRD.doc.pdf), October 2007.

Jaffe, Greg. "Marching Orders: To Keep Recruits, Boot Camp Gets a Gentle Revamp; Army Offers More Support, Sleep, Second Helpings; Drill Sergeants' Worries; 'It Would Look So Much Nicer.' " *Wall Street Journal,* February 15, 2006.

Kennedy, Kelly. "Sleep Starved." *Army Times,* May 19, 2006.

Khatchadourian, Raffi. "The Kill Company." *New Yorker,* July 6, 2009.

Killgore, William D., Arthur Estrada, Tiffany Rouse, Robert M. Wildzunas, and Thomas J. Balkin. *Sleep and Performance Measures in Soldiers Undergoing Military Relevant Training.* Fort Rucker, AL: U.S. Army Aeromedical Research Laboratory, Warfighter Performance and Health Division, June 2009.

Killgore, William D. S., Sharon A. McBride, Desiree B. Killgore, and Thomas J.

Balkin. "The Effects of Caffeine, Dextroamphetamine, and Modafinil on Humor Appreciation During Sleep Deprivation." *Sleep*, vol. 29 (June 2006).

Kushida, Clete A. *Sleep Deprivation: Basic Science, Physiology and Behavior.* New York: Marcel Dekker, 2005.

Laurence, Charles. "Ready for War in 2005: The Soldier Who Never Sleeps." *Daily Telegraph*, January 5, 2003.

Lehrer, Jonah. *How We Decide.* New York: Houghton Mifflin Harcourt, 2009.

Martz, Ron. "War Story: GI Joe." *Atlanta Magazine*, March 2008.

Mestrovic, Stjepan. *The Good Soldier on Trial: A Sociological Study of Misconduct by the US Military Pertaining to Operation Iron Triangle, Iraq.* New York: Algora, 2009.

Miller, N. L., and R. Firehammer. "Avoiding a Second Hollow Force: The Case for Including Crew Endurance Factors in the Afloat Staffing Policies of the U.S. Navy." *Naval Engineers Journal*, vol. 119, no. 1 (2007).

Miller, N. L., P. Matsangas, and L. G. Shattuck. "Fatigue and Its Effect on Performance in Military Environments," in P. A. Hancock and J. L. Szalma, eds., *Performance under Stress.* Burlington, VT: Ashgate, 2007.

Miller, Nita Lewis, and Lt. John Nguyen. "Working the Nightshift on the USS John C. Stennis: Implications for Enhancing Warfighter Effectiveness." Conference paper, Human Systems Integration Symposium, Vienna, VA, May 1, 2003.

Robson, Seth. "In Video, Leahy Tells of Shooting Iraqi Detainees." *Stars and Stripes*, February 20, 2009.

Robson, Seth. "Report: Troops Need More Sleep." *Stars and Stripes*, March 17, 2009.

Scott, William B. "Crew Fatigue Emerging as Critical Safety Issue." *Aviation Week and Space Technology*, April 8, 1996.

Shanker, Thom, with Mary Duenwald. " 'Go Pills' Center-Stage at U.S. Pilots' Hearing: Effect of Amphetamine Use Is Murky." *International Herald Tribune*, January 20, 2003.

Shay, Jonathan. "Ethical Standing for Commander Self-Care: The Need for

Sleep," *Parameters* (U.S. Army War College), vol. 28, no. 2 (Summer 1998).

Smith, Elliot Blair. "Fatigue a Formidable Enemy within the Ranks: Sleep Deprivation Taking Toll on Troops, So Weary Warriors Catch Catnaps When and Where They Can." *USA Today,* March 28, 2003.

Squeo, Anne Marie, and Nicholas Kulish. "A Growing Threat to Troops in Iraq: Sleep Deprivation." *Wall Street Journal,* March 27, 2003.

Von Zielbauer, Paul. "Court Papers Describe Killings of Prisoners by Three U.S. Troops in Iraq." *International Herald Tribune,* August 28, 2008.

Von Zielbauer, Paul. "U.S. Soldiers Executed Iraqis, Statements Say." *New York Times,* August 26, 2008.

Wesensten, Nancy J., Gregory Belenky, and Thomas J. Balkin. "Sleep Loss: Implications for Operational Effectiveness and Current Solutions," in Thomas W. Britt, Carl A. Castro, and Amy B. Adler, eds., *Military Life: The Psychology of Serving in Peace and Combat,* vol. 1. Westport, CT: Praeger Security International, 2005.

Chapter 8: Bumps in the Night

Bachelder, Vance, and Michel A. Cramer Bornemann. "New Research in Sleep-Disorder Breathing." *RT: For Decision Makers in Respiratory Care* (June 2003).

Callwood, June. *The Sleepwalker.* Toronto: Lester and Orpen Dennys, 1990.

Cartwright, Rosalind. "Sleepwalking Violence: A Sleep Disorder, a Legal Dilemma, and a Psychological Challenge." *American Journal of Psychiatry,* vol. 161, no. 7 (July 2004).

Cramer Bornemann, Michel A., Mark W. Mahowald, and Carlos H. Schenck. "Parasomnias: Clinical Features and Forensic Implications." *Chest,* vol. 130, no. 2 (August 2006).

Denno, Deborah W. "Crime and Consciousness: Science and Involuntary Acts." *Minnesota Law Review,* vol. 87 (2002).

Denno, Deborah W. "Criminal Law in a Post-Freudian World." *University of Illinois Law Review*, vol. 601 (2005).

Denno, Deborah W. "A Mind to Blame: New Views on Involuntary Acts." *Behavioral Sciences and the Law*, vol. 21 (2003).

Krasnowski, Matt. "Sleepwalking Defense Is Called 'Sophistry'; Killer Gets 26 Years." *San Diego Union Tribune*, August 20, 2004.

Lauerma, Hannu. "Fear of Suicide during Sleepwalking." *Psychiatry*, vol. 59, no. 2 (Summer 1996).

Mahowald, M. W., C. H. Schenck, M. Goldner, V. Bachelder, and M. Cramer-Bornemann. "Parasomnia Pseudo-Suicide." *Journal of Forensic Sciences*, vol. 48, no. 5 (2003).

Mahowald, Mark W., and Carlos H. Schenck. "Parasomnias: Sleepwalking and the Law." *Sleep Medicine Reviews*, vol. 4, no. 4 (2000).

"Man Acquitted of Sleepwalking Murder Running for School Trustee in Durham." CityNews.ca. October 27, 2006.

McLeod, Keith. "A Decent Man and Devoted Husband; Dad Who Strangled Wife in Sleep Is Cleared." *Daily Record* (Glasgow), November 21, 2009.

Milliet, Nicolaa, and Wolfgang Ummenhofer. "Somnambulism and Trauma: Case Report and Short Review of the Literature." *Journal of Trauma: Injury, Infection, and Critical Care*, vol. 47 (August 1999).

Morse, Stephen J., and Morris B. Hoffman. "The Uneasy Entente between Legal Insanity and Mens Rea: Beyond Clark V. Arizona." *Journal of Criminal Law and Criminology*, vol. 97, no. 4 (Summer 2007).

Schenck, C. H., I. Arnulf, and M. W. Mahowald. "Sleep and Sex: What Can Go Wrong? A Review of the Literature on Sleep Related Disorders and Abnormal Sexual Behaviors and Experiences." *Sleep*, vol. 30 (June 2007).

Schenck, Carlos H., Samuel Adams Lee, Michel A. Cramer Bornemann, and Mark W. Mahowald. "Potentially Lethal Behaviors Associated with Rapid Eye Movement Sleep Behavior Disorder: Review of the Literature and Forensic Implications." *Journal of Forensic Sciences*, vol. 54, no. 6 (2008).

Sleep Runners: The Stories behind Everyday Parasomnias. Brian L. Dehler, dir. Documentary. Slow-Wave Films, 2011.

Stryker, Jeff. "Sleepstabbing: The Strange Science of Sleep Behavior and One Verdict: Guilty!" *Salon*, July 8, 1999.

Tighe, Janet A., and Francis Wharton. "The Nineteenth-Century Insanity Defense: The Origins of a Reform Tradition." *American Journal of Legal History*, vol. 27, no. 3 (July 1983).

Vienneau, David. "Sleepwalk Murder Acquittal Upheld by Supreme Court." *Toronto Star*, August 27, 1992.

Chapter 9: Game Time

Bronson, Po. "Snooze or Lose. " *New York*, October 7, 2007.

Brown, Frederick M., Evan E. Neft, and Cynthia M. LaJambe. "Collegiate Rowing Crew Performance Varies by Morningness-Eveningness." *Journal of Strength and Conditioning Research*, vol. 22, no. 6 (November 2008).

Corbett Dooren, Jennifer. "Later Start to School Boosts Teens' Health." *Wall Street Journal*, July 6, 2010.

Dahl, R. E., and A. G. Harvey. "Sleep in Children and Adolescents with Behavioral and Emotional Disorders." *Sleep Medicine Clinics*, vol. 2 (September 2007).

Doskoch, Peter. "Putting Time on Your Side." *Psychology Today*, vol. 30, no. 2 (March/April 1997).

Frias, Carlos. "Baseball and Amphetamines." *Palm Beach Post*, April 2, 2006.

Gangwisch, James E., Lindsay A. Babiss, Dolore Malaspina, J. Blake Turner, Gary K. Zammit, and Kelly Posner. "Earlier Parental Set Bedtimes as a Protective Factor against Depression and Suicidal Ideation." *Sleep*, vol. 33 (January 2010).

Mah, Cheri. "Extended Sleep and the Effects on Mood and Athletic Performance in Collegiate Swimmers." Presented at the 2008 annual meeting of the Associated Professional Sleep Societies, Baltimore, MD, June 9, 2008.

O'Brien, Louise M., Neali H. Lucas, Barbara T. Felt, Timothy F. Hoban, Deborah L. Ruzicka, Ruth Jordan, Kenneth Guire, and Ronald D. Chervin. "Aggressive Behavior, Bullying, Snoring, and Sleepiness in Schoolchildren." *Sleep Medicine*, Vol. 12, no. 7 (August 2011).

Postolache, T. T., T. M. Hung, R. N. Rosenthal, J. J. Soriano, F. Montes, and J. W. Stiller. "Sports Chronobiology Consultation: From the Lab to the Arena." *Clinical Sports Medicine*, vol. 24 (April 2005).

Postolache, Teodor T., and Dan A. Orenc. "Circadian Phase Shifting, Alerting, and Antidepressant Effects of Bright Light Treatment." *Clinical Sports Medicine*, vol. 24 (April 2005).

Reilly, Thomas. "The Body Clock and Athletic Performance." *Biological Rhythm Research*, vol. 40, no. 1 (February 2009).

Rosbash, Michael. "A Biological Clock." *Daedalus*, vol. 132, no. 2 (Spring 2003).

Samuels, Charles. "Sleep, Recovery, and Performance: The New Frontier in High-Performance Athletics." *Neurological Clinics*, vol. 26 (February 2008).

Smith, Roger S., Christian Guilleminault, and Bradley Efron. "Sports, Sleep and Circadian Rhythms: Circadian Rhythms and Enhanced Athletic Performance in the National Football League." *Sleep*, vol. 20 (May 1997).

Stein, Jeannine. "Athletes Who Sleep More May Score More." *Los Angeles Times*, June 18, 2007.

Travis, John. "Does March Madness Need a Time-Out?" *Science News*, vol. 156, no. 19 (November 1999).

Tucker, Jill. "Sleep May Limit Teen's Depression." *San Francisco Chronicle*, November 13, 2009.

Tyack, David, and Larry Cuban. *Tinkering toward Utopia: A Century of Public School Reform*. Cambridge, MA: Harvard University Press, 1995.

Wahlstrom, K. "Accommodating the Sleep Patterns of Adolescents within Current Educational Structures: An Uncharted Path," in M. Carskadon, ed., *Adolescent Sleep Patterns: Biological, Social, and Psychological Influences.* New York: Cambridge University Press, 2002.

Wahlstrom, K. L. "The Prickly Politics of School Starting Times." *Kappan,* vol. 80, no. 5 (1999).

Wahlstrom, Kyla. "Changing Times: Findings from the First Longitudinal Study of Later High School Start Times." *NASSP Bulletin,* vol. 86, no. 633 (December 2002).

Chapter 10: Breathe Easy

Aleccia, JoNel. "Heavy, Drowsy Truckers Pose Risk on the Road." *MSNBC.com,* June 14, 2009.

American Sleep Apnea Association. "Apnea Support Forum." http://www.apneasupport.org/help-1st-sleep-study-last-night-scared-angry-sad-dismayed-t26555.html. Accessed August 2011.

"A Brief History of OSA." *ResMedica Clinical Newsletter,* no. 14 (2011).

Davidson, Terence M. "The Great Leap Forward: The Anatomic Basis for the Acquisition of Speech and Obstructive Sleep Apnea." *Sleep Medicine,* vol. 4 (May 2003).

Dement, William C., and Christopher Vaughan. *The Promise of Sleep: A Pioneer in Sleep Medicine Explores the Vital Connection between Health, Happiness, and a Good Night's Sleep.* New York: Delacorte Press, 1999.

Diamond, J. *The Third Chimpanzee: The Evolution and Future of the Human Animal.* New York: HarperCollins, 1992.

Durand, G., and S. N. Kales. "Obstructive Sleep Apnea Screening during Commercial Driver Medical Examinations: A Survey of ACOEM Members." *Journal of Occupational and Environmental Medicine,* vol. 51 (October 2009).

Espie, Colin A., and Niall M. Broomfield. "The Attention–Intention–Effort Pathway in the Development of Psychophysiologic Insomnia: A Theoretical Review." *Sleep Medicine Reviews*, vol. 10, no 4 (August 2006).

Government Accountability Office. *Commercial Drivers: Certification Process for Drivers with Serious Medical Conditions.* Report no. GAO-08-1030T. Washington, DC: Government Accountability Office, July 24, 2008.

Isono, Shiroh, John E. Remmers, Atsuko Tanaka, Yasuhide Sho, Jiro Sato, and Takashi Nishino. "Anatomy of Pharynx in Patients with Obstructive Sleep Apnea and in Normal Subjects." *Journal of Applied Physiology*, vol. 82, no. 4 (April 1997).

Kirby, Tony. "Colin Sullivan: Inventive Pioneer of Sleep Medicine." *Lancet*, vol. 377 (April 2011).

Kumar, R., B. V. Birrer, P. M. Macey, M. A. Woo, R. K. Gupta, F. L. Yan-Go, and R. M. Harper. "Reduced Mammillary Body Volume in Patients with Obstructive Sleep Apnea." *Neuroscience Letters*, vol. 438 (June 2008).

Macey, P. M., R. Kumar, M. A. Woo, E. M. Valladares, F. L. Yan-Go, and R. M. Harper. "Brain Structural Changes in Obstructive Sleep Apnea." *Sleep*, vol. 31 (July 2008).

Maher, Kris. "The New Face of Sleep—As Patients Balk at Bulky Masks, New Efforts to Treat Sleep Apnea." *Wall Street Journal*. February 2, 2010.

National Commission on Sleep Disorders Research. *Report of the National Commission on Sleep Disorders Research.* Washington, DC: U.S. Government Printing Office, 1992.

Parks, P. D., G. Durand, A. J. Tsismenakis, A. Vela-Bueno, and S. N. Kales. "Screening for Obstructive Sleep Apnea during Commercial Driver Medical Examinations." *Journal of Occupational and Environmental Medicine*, vol. 51 (October 2009).

"ResMed Inc. Announces Record Financial Results for the Quarter and Twelve Months Ended June 30, 2010." ResMed News Release, August 5, 2010.

Yaffe, Kristine, Alison M. Laffan, Stephanie Litwack Harrison, Susan Redline, Adam P. Spira, Kristine E. Ensrud, Sonia Ancoli-Israel, and Katie

L. Stone. "Sleep-Disordered Breathing, Hypoxia, and Risk of Mild Cognitive Impairment and Dementia in Older Women." *JAMA*, vol. 306 (August 2011).

Chapter 11: Counting Sheep

Alderman, Lesley. "Cost-Effective Ways to Fight Insomnia." *New York Times*, June 5, 2009.

Ansfield, M. E., D. M. Wegner, and R. Bowser. "Ironic Effects of Sleep Urgency." *Behaviour Research and Therapy*, vol. 34 (July 1996).

Armstrong, David. "Sales of Sleeping Pills Are Seeing a Revival; Lunesta's Big Launch." *Wall Street Journal*, April 19, 2005.

Edwards, Jim. "Lunesta's Hit Marketing May Be a Dying Breed." *Brandweek*, August 22–29, 2005.

Edwards, Jim. "Sleep: Perchance to Dream." *Brandweek*, October 9, 2006.

Fratello, F. "Can an Inert Sleeping Pill Affect Sleep? Effects on Polysomnographic, Behavioral and Subjective Measures." *Psychopharmacology*, vol. 181 (October 2006).

Herzberg, David. *Happy Pills in America: From Miltown to Prozac*. Baltimore: John Hopkins University Press, 2009.

Institute of Medicine. *Sleeping Pills, Insomnia and Medical Practice*. Washington, DC: National Academy of Sciences, 1979.

Martin, Emily. "Sleepless in America," in Janis H. Jenkins, ed., *Pharmaceutical Self: The Global Shaping of Experience in an Age of Psychopharamacology*. Santa Fe, NM: SAR Press, 2011.

Morin, Charles. "Sequential Treatment for Chronic Insomnia: A Pilot Study." *Behavioral Sleep Medicine*, vol. 2, no. 2 (2004).

Morin, Charles, Célyne Bastien, Bernard Guay, Monelly Radouco-Thomas, Jacinthe Leblanc, and Annie Vallières. "Cognitive Behavioral Therapy, Singly and Combined with Medication, for Persistent Insomnia." *JAMA*, vol. 301 (May 2009).

Morin, Charles, Célyne Bastien, Bernard Guay, Monelly Radouco-Thomas, Jacinthe Leblanc, and Annie Vallières. "Randomized Clinical Trial of Supervised Tapering and Cognitive Behavior Therapy to Facilitate Benzodiazepine Discontinuation in Older Adults with Chronic Insomnia." *American Journal of Psychiatry*, vol. 161 (2004).

Morin, Charles, Annie Vallières, and Hans Ivers. "Dysfunctional Beliefs and Attitudes about Sleep." *Sleep*, vol. 30 (November 2007).

Morin, Charles M., Cheryl Colecchi, Jackie Stone, Rakesh Sood, and Douglas Brink. "Behavioral and Pharmacological Therapies for Late-Life Insomnia." *JAMA*, vol. 281 (March 17, 1999).

Morris, H. H., and M. L. Estes. "Traveler's Amnesia. Transient Global Amnesia Secondary to Triazolam," *JAMA*, vol. 258 (August 1987).

Murphy, Shelley. "Unruly Jet Passenger Pleads Guilty." *Boston Globe*, September 8, 2005.

National Sleep Foundation. "2003 Sleep in America Poll." Prepared by WB&A Market Research. Found at http://www.sleepfoundation.org/sites/default/files/2003SleepPollExecSumm.pdf.

Roth, Thomas. "Insomnia: Definition, Prevalence, Etiology, and Consequences." *Journal of Clinical Sleep Medicine*, vol. 3 (August 15, 2007).

Saul, Stephanie. "Sleep Drugs Found Only Mildly Effective, but Wildly Popular." *New York Times*, October 23, 2007.

Tsai, M. J., Y. H. Tsai, and Y. B. Huang. "Compulsive Activity and Anterograde Amnesia after Zolpidem Use." *Clinical Toxicology*, vol. 45, no. 2 (2007).

Wegner, D. M., and J. W. Pennebaker, eds. *Handbook of Mental Control*. Englewood Cliffs, NJ: Prentice-Hall, 1993.

Worthman, C. M., and M. Melby. "Toward a Comparative Developmental Ecology of Human Sleep," in M.A. Carskadon, ed., *Adolescent Sleep Patterns: Biological, Social, and Psychological Influences*. New York: Cambridge University Press, 2002.

Chapter 12: Mr. Sandman

"Air India Plane Crash: 'Sleepy' Pilot Blamed." BBC News, November 17, 2010.

Bader, G. G., and S. Engdal. "The Influence of Bed Firmness on Sleep Quality." *Applied Ergonomics,* vol. 31 (October 2000).

Bergholdt, K. "Better Backs by Better Beds?" *Spine,* vol. 33 (April 2008).

Calamari, Luigi. "Effect of Different Free Stall Surfaces on Behavioural, Productive and Metabolic Parameters in Dairy Cows." *Applied Animal Behaviour Science,* vol. 120 (August 2009).

Dement, William C., and Christopher Vaughan. *The Promise of Sleep: A Pioneer in Sleep Medicine Explores the Vital Connection between Health, Happiness, and a Good Night's Sleep.* New York: Delacorte Press, 1999.

"Directions for the Management of Sleep." *Dublin Penny Journal,* vol. 2, no. 74 (1833).

Gerber, Markus, Serge Brand, and Edith Holsboer-Trachsler. "Fitness and Exercise as Correlates of Sleep Complaints: Is It All in Our Minds?" *Medicine and Science in Sports and Exercise (Journal of the American College of Sports Medicine),* vol. 42 (May 2010).

Khalsa, Sat Bir S. "Treatment of Chronic Insomnia with Yoga: A Preliminary Study with Sleep–Wake Diaries." *Applied Psychophysiology and Biofeedback,* vol. 29 (December 2004).

Lack, L. C., M. Gradisar, E. J. Van Someren, H. R. Wright, and K. Lushington. "The Relationship between Insomnia and Body Temperatures." *Sleep Medicine Review,* vol. 12 (August 2008).

Leea, Hyunj, and Sejin Park. "Quantitative Effects of Mattress Types (Comfortable vs. Uncomfortable) on Sleep Quality through Polysomnography and Skin Temperature." *International Journal of Industrial Ergonomics,* vol. 36 (November 2006).

Mooallem, Jon. "The Sleep Industrial Complex." *New York Times,* November 18, 2007.

Onen, S. H., F. Onen, D. Bailly, and P. Parquet. "Prevention and Treatment of Sleep Disorders through Regulation of Sleeping Habits." *Presse Med*, vol. 23 (March 1994).

Reid, K. J., K. G. Baron, B. Lu, E. Naylor, L. Wolfe, and P. C. Zee. "Aerobic Exercise Improves Self-Reported Sleep and Quality of Life in Older Adults with Insomnia." *Sleep Medicine*, vol. 11 (October 2010).

Singh, Nalin, Theodora M. Stavrinos, Yvonne Scarbek, Garry Galambos, Cas Liber, and Maria A. Fiatarone Sing. "A Randomized Controlled Trial of High versus Low Intensity Weight Training versus General Practitioner Care for Clinical Depression in Older Adults." *Journal of Gerontology: Biological Sciences*, vol. 60 (June 2005).

Tworoger, Shelley S., Yutaka Yasui, Michael V. Vitiello, and Robert S. Schwartz. "Effects of a Yearlong Moderate-Intensity Exercise and a Stretching Intervention on Sleep Quality in Postmenopausal Women." *Sleep*, vol. 26 (November 2003).

Youngstedt, Shawn, and Christopher Kline. "Epidemiology of Exercise and Sleep." *Sleep and Biological Rhythms*, vol. 4 (October 2006).